Comparators

Michael C. Fischer

Comparators

 Springer

Michael C. Fischer
Fischer Consulting
Palo Alto, CA, USA

ISBN 978-3-030-95744-5   ISBN 978-3-030-95742-1   (eBook)
https://doi.org/10.1007/978-3-030-95742-1

This Springer imprint is published by the registered company Springer Nature Switzerland AG
The registered company address is: Gewerbestrasse 11, 6330 Cham, Switzerland

Preface

If you are faced with doing a circuit or system design where one of the blocks in the block diagram (or many of the blocks) is a comparator, and you have looked for guidance in the details of that task from the manufacturers of comparators, or the textbooks that mention comparators, and still feel uninformed, that may be why you picked up this book. It is my intention that I have left no stone unturned in addressing each and every issue that can affect the engineering tasks related to comparators, at least from the viewpoint of how their performance can affect the system in which they are a critical component.

There are two main thrusts of this material: one is the specification and performance of the comparator as an element, and the other is the circuitry surrounding and supporting the comparator. The data sheets for comparators are missing some information that can be critically important, as will be analyzed in the text. The topology of the circuitry around comparators can have many forms, some more effective than others, with the better ones covered here.

The impetus for this book came from a system design project for a consulting client that included a stage best served by a comparator. Being familiar with comparators, I began the process of choosing a suitable device. I wanted the chosen comparator to function with its supporting circuitry in an optimal way, and therefore set about searching the literature for design guidance. I was shocked by the dearth of circuit topology and quantitative design criteria. That left me estimating parameters that I would have preferred to deal with in a more definitive fashion. My viewpoint is more that of a user of comparator products rather than a designer of the internals of a comparator, and that guides the approach of this book.

With gratitude, I acknowledge the encouragement and support of Marilyn Roberts, without whom this work would have taken longer and been more difficult. I dedicate this book to her.

Palo Alto, CA Michael Fischer

About the Author

The author is an independent consultant/author specializing in system architecture, circuit development, and board-level hardware design and analysis with over 25 years of experience in electronics engineering covering instrumentation, consumer, industrial, electro-optical, data storage, telecommunications, navigation, and frequency control. He has 13 patents issued, with 3 applications pending. Since 2001, he has been the owner and principal consultant at Fischer Consulting (R). Previously, he worked at Hewlett-Packard for 28 years, Magnavox Research Labs for 3 years, Tracor Inc. for 6 years, and the Radar Division of the Defense Research Lab at the University of Texas, Austin, for 3 years. He holds a BSEE from the University of Texas at Austin and memberships in IEEE as a life senior, Professional and Technical Consultants Association, Consultants' Network of Silicon Valley, and Audio Engineering Society.

About This Book

Circuit designs that accomplish the conversion of an analog signal to a digital signal of a single bit are explored. Starting with the simple comparator, many alternative circuit arrangements and enhancements are elaborated, including hysteresis, negative feedback, and a variety of adaptive thresholds. Further, the nonideal behavior of practical elements and circuits is covered, including input offsets, noise, delay, delay dispersion, and oscillation, along with techniques for dealing with these aspects. The wide variety of available components is discussed in terms of performance and applicability.

List of Symbols

BW	Bandwidth in hertz
f	Frequency in hertz
f_o	Frequency in hertz of the main signal in question
f_{so}	Frequency of self-oscillation, Hz
L	Single-sideband phase noise, dBc/Hz
L_r	Single-sideband phase noise, power ratio per Hz of bandwidth
Q	Quality factor of a resonant circuit
SNR	Signal-to-noise ratio
t_r	Risetime of a waveform
t_{pd}	Propagation delay time of a comparator
V_{be}	Base-emitter voltage
V_{cc}	Positive supply voltage
V_i	Noise voltage, V rms, at the input of a comparator
V_n	Noise voltage, V rms
V_{oh}	Comparator output voltage, high state
V_{ol}	Comparator output voltage, low state
V_{osh}	Offset voltage effect to hysteresis
V_{pk}	Volts peak, amplitude of a signal
V_{pp}	Volts peak-to-peak, amplitude of a signal
V_{ref}	Reference voltage input to the comparator
V_{sig}	Volts peak-to-peak, amplitude of a signal input to a comparator
V_{test}	Volts peak, amplitude of a test signal input to a comparator
φ	Phase variation, radians rms
π	3.14159...
τ	Time jitter, seconds rms; or period of a waveform, seconds

Acronyms

8PSK	Eight-ary phase-shift keying
AC	Alternating current
ADC	Analog-to-digital converter
AGC	Automatic gain control
AM-to-PM	Amplitude modulation-to-phase modulation
BJT	Bipolar junction transistor
BPSK	Binary phase-shift keying
CMOS	Complementary metal-oxide semiconductor
DAC	Digital-to-analog convertor
dB	Decibel
DC	Direct current
DFT	Discrete Fourier transform
DSP	Digital signal processing
ECL	Emitter-coupled logic
FET	Field effect transistor
FFT	Fast Fourier transform
FM	Frequency modulation
IC	Integrated circuit
IF	Intermediate frequency
IPI	Inter-pulse interval
JFET	Junction field effect transistor
LED	Light-emitting diode
MOS	Metal-oxide semiconductor
MOSFET	Metal-oxide semiconductor field effect transistor
NECL	Negative emitter-coupled logic
NMOS	Negative metal-oxide semiconductor
OpAmp	Operational amplifier
PECL	Positive emitter-coupled logic
PFM	Pulse frequency modulation
PM	Phase modulation
PMOS	Positive metal-oxide semiconductor
PLL	Phase-locked loop
PSK	Phase-shift keying
PWM	Pulse width modulation
QPSK	Quadrature phase-shift keying
RC	Resistance-capacitance
RF	Radio frequency
rms	Root mean square
RS	Reset-set (flip-flop)
SNR	Signal-to-noise ratio
TTL	Transistor-transistor logic
VCO	Voltage-controlled oscillator
VFC	Voltage-to-frequency converter
VU	Volume unit—audio-level measure

Contents

Chapter 1
Basic Concepts

1.1 Introduction

Wherever an analog signal carrying a stream of digital information must be interpreted to become a digital data stream, with unambiguous levels of "ones" and "zeroes," a decision-making circuit must be arranged to provide this function. The basic comparator supplies this function, but depending on vagaries of the incoming signal, there are several parameters to be considered, and a number of circuit enhancement alternatives. This publication explores these topics and offers techniques of analysis and circuit topologies to cover a wide range of needs.

As more and more signal processing is done in digital hardware, the comparator is the canonical element that serves the process of converting the analog signals of the real world to bitstreams so that they can be processed digitally.

1.2 Synonyms

Comparator, data slicer, hard decision circuit, hard limiter, one-bit analog-to-digital converter, trigger, Schmitt trigger, and zero-crossing detector are all essentially synonymous terms for the same function.

1.3 Basic Functions

Data streams typically come out of analog circuitry, whether the demodulator of a radio receiver, an optical fiber receiver, or data channel preamplifier of a rotating storage device—hard drive, optical drive, or magneto-optical drive. At this point in all these and other systems, the analog variations in voltage must be compared with

M. C. Fischer, *Comparators*, https://doi.org/10.1007/978-3-030-95742-1_1

Fig. 1.1 Basic comparator

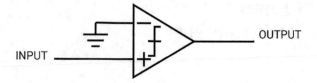

a threshold voltage and a hard decision made as to whether the present signal level represents a "one" or a "zero." In the simplest case the comparator function serves the need, as in Fig. 1.1.

A familiar application for the function of a comparator is the "idiot light" on the dashboard of an automobile that comes on to alert to the condition of low oil pressure in the engine. The oil pressure is an analog quantity, and a comparator makes a decision to turn on the warning lamp when the oil pressure drops below a preset threshold.

The fundamental circuit function that we are dealing with, the comparator, and the element that performs this function, is an element with analog input (or differential pair of inputs) and digital output. The function performed is to compare the input level with a reference level that is either internal or present at the other input of a differential pair of inputs. Whether the second input is an internal reference or an actual input terminal, we need to name the two inputs to describe the function of the element. The most widely used names are "inverting" and "non-inverting" for the two inputs.

The names of the inputs follow their definitions. If the voltage at the non-inverting input is greater than the inverting input, then the result of the comparison function is an output that is a logic high level, "1," hence the choice of the name "non-inverting." The other result of the comparison is an output that is a logic low level, "0," if the inverting input is greater than the non-inverting input. Consistent with these definitions are the labels "+" for the non-inverting and "−" for the inverting inputs.

In summary, then, a comparator has two analog voltage inputs, one inverting and one non-inverting, and one digital output (sometimes a differential pair of digital outputs), where one of the two inputs may be an internal reference level. Calling the inputs "voltage" is to imply that they are high impedance, at least relatively high. The output is generally specified as a logic level, often corresponding to the levels of one or more logic families. The output is generally low impedance, or sometimes open collector or drain, or the output may be a differential pair.

The comparator symbol often contains a representation of the transfer function as shown in Fig. 1.1, to distinguish it from a standard operational amplifier. In subsequent figures, we will omit the transfer function symbol to simplify the graphics.

The terms "threshold level," "decision level," and "reference level" are used in this discussion more or less interchangeably. "Decision level" means specifically the voltage at the comparator input terminal where the output changes state. "Reference level" has the slightly different meaning as the voltage applied to the comparator reference input intending that it become the decision level. From these points one may see that the difference between "decision level" and "reference level" is the offset voltage of the comparator. "Threshold level" refers to the concept of comparator operation and is more closely associated with "decision level." In Fig. 1.1 above, the reference level is circuit ground.

The decision level may be fixed, adaptive, or programmed. A fixed decision level is simply a constant dc voltage chosen to make the best choice between "high" and "low." An adaptive decision level changes in some automatic way to adjust to changes in the signal plus interference, plus dc drifts. A programmed decision level may be set by the output of a digital to analog converter (DAC) or some analog control signal.

The typical functioning of a comparator may be viewed graphically in the figure below. Consider the three panels to be an oscilloscope image of three channels of voltage versus time. The top panel is a portion of a data stream to be transmitted over a communication link, the bit pattern is:

```
0 1 0 0 1 0 1 1 0 0
```

After this data stream is band-limited by transmission through a communications channel, the recovered signal at the receiving end in its analog form is shown in the middle panel. The delay is due to the group delay of the band limiting filtering as well as the channel propagation delay. In this case almost all the delay is due to the filtering.

This received analog signal in the middle panel must be presented to a comparator to re-establish the digital levels corresponding to the original data stream, though delayed. The result of the comparator's decisions are shown in the third panel.

In Fig. 1.2, the data bits have a duration of 10 µs, corresponding to the data rate of 100 kb/s, and the group delay of the band-limiting filter is about 8 µs. The band-limiting filter is a three-pole, near linear phase (constant group delay) form having a −3 decibel (dB) corner at 27 kHz, and is about −8 dB at half the data rate. Half the data rate is the frequency of the highest fundamental component of the data spectrum, present when the data pattern is alternating ones and zeros.

It is clear that the threshold level or decision level of this comparator is at 1.3 V with little or no hysteresis. The slew rate of the input to the comparator, at the decision level, is about 0.2 V/µs. Slew rate and hysteresis will be discussed in more detail later.

The fact that the recovered data stream transitions occur at times skewed by a few microseconds in addition to the delay is evidence of the inter-symbol interference that occurs in such a channel.

1.3.1   Comparator Versus Operational Amplifier

The comparator function is similar to that of an operational amplifier; high-input impedance, low-output impedance, high gain, low-input offset voltage, and low bias currents. However, the usual operational amplifier is less desirable for this function

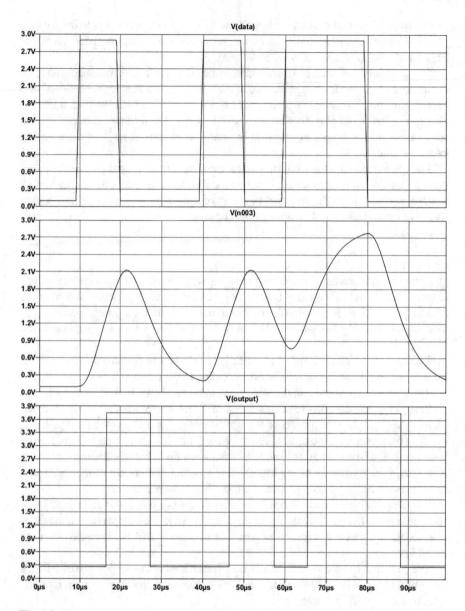

Fig. 1.2 Comparator action in a communications channel

than a comparator specifically designed for the purpose. The comparator function requires that the output change quickly from the high to the low overload states and back again, with minimal delays. Operational amplifiers are not optimized for overload recovery, while comparators are. The circuitry following a comparator will be logic circuits, and operational amplifier outputs are not characterized or specified for driving logic levels while comparators are. The propagation delay through this stage is usually of concern, especially if it is not constant. Operational amplifiers are not specified as to propagation delay or to changes in overload recovery delay versus input overdrive changes, while the better comparators are.

An operational amplifier (op-amp) may precede a comparator to enhance the operation of the comparator. This is discussed later in more detail in the section Op-Amp Enhanced Comparator.

While keeping all these issues in mind, in a system design where there is an extra unused opamp, and a low-performance comparator is needed, it may be reasonable to use an opamp as a comparator.

Reference: Trump (2012)

1.3.2   Functional Details

The operation of the circuit in Fig. 1.1 is that the output is a logic level that is determined as high or low by the input signal being above or below the threshold reference voltage, respectively. In Fig. 1.1, the threshold reference voltage is zero or ground, shown connected to the inverting input of the comparator, but any other voltage could be connected here to serve as the threshold reference. The convention of zero reference voltage will be maintained to simplify the presentation of many of the circuits to follow. Other implementations are workable, for instance, in a single-supply environment, a virtual (signal) ground bus of half the supply voltage may be generated and used as the "ground" reference.

An inverting comparator simply interchanges the input and the threshold reference connections from those shown in Fig. 1.1. This causes the output to go to its high level when the input is below the threshold and vice versa. Inverting and non-inverting connections are equally useful and are typically chosen to suit the needs of the surrounding circuitry. In some of the enhancements that follow, the choice of inverting or not is constrained by other circuit considerations, namely hysteresis, and typically the logic circuitry that follows the comparator stage can be designed to accommodate either sense, at worst by the addition of an inverter stage.

The simplest basic comparator structure is a pair of transistors (preferably matched), either field-effect transistor (FET) or bipolar, connected as in Fig. 1.3. This arrangement is sometimes called an "emitter coupled differential amplifier," or a "long-tailed pair."

Integrated circuit comparators offer specified performance, that is much better, for a cost that is similar, versus comparators made from discrete components. For that reason, this document will concentrate on them instead of discrete component implementations. In the circuit topologies to follow, the power connections to the integrated circuit (IC) comparators are not shown on the schematics to emphasize signal paths and functionality. Data sheet power recommendations should be followed, especially including power bypass capacitors located as near the IC as possible in every case, and not shared with any other IC. Further discussions of these

Fig. 1.3 Basic comparator
circuit

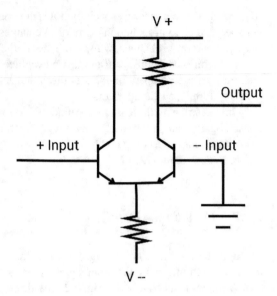

and other points are covered in Horowitz & Hill (2015), Malmstadt, Enke & Crouch (1981) and the Wikipedia (2020) article.

To aid in understanding the details of what might occur in an actual comparator circuit, the following figure and description deal with the product family that includes the LM339 from National Semiconductor, now Texas Instruments, products that are also available from a number of manufacturers.

Fig. 1.4 represents a family of quad comparators has several broadly useful features and a few limitations. They can be operated over a supply voltage range of 2–32 V, and besides the quad version, are offered in dual, LM393-B, and single, TL331B in an SOT-23-5 surface mount package. These parts are very widely available, and for the quad, the price can range as low as $0.06 in 1000 quantity, which amounts 1.5 cents per comparator. The input common mode voltage range is from 0 to V_{cc}-1.5 V. The input offset voltage is typically 1 mV and can be limited to 4 mV maximum over temperature. The input offset current is typically 3 nA and can be limited to 100 nA over temperature. The input bias current is typically −25 nA and can be limited to −300 nA over temperature. These tighter limits are offered in more demanding specifications on certain devices. The output can drive a voltage much different from the V_{cc}, higher or lower, and being open collector, allows for connecting together the outputs of several comparators to a common load, thus providing a wired-OR function. The output can sink 4 mA with an output voltage of typically 150 mV, and no more than 700 mV over temperature. The total supply current for four comparators is typically 0.8 mA and a maximum of 2 mA at 25 °C. They are not fast, having a propagation delay time of 300 ns, and a rather poor dispersion of 21 ns/dB of input signal level change (defined and discussed later). There is no internal hysteresis. The operation of the circuit in Fig. 1.4 is described in the following excerpt from the reference Russell (1972):

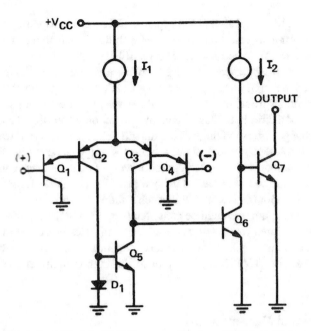

Fig. 1.4 Simplified circuitry of the LM339, one section, from IEEE J. Solid-State Circuits, vol. SC-7, pp. 446–454, Dec. 1972. (With permission from the IEEE)

The input stage "… employs p-n-p rather than n-p-n transistors in the differential amplifier." The maximum voltage on the collectors of Q2 and Q3 in the differential amplifier is one base-emitter voltage drop above ground due to the diode D1 in the differential-to-single-ended converter and the V_{be} of transistor Q6. This allows the dc level shift from base to emitter of the substrate p-n-p followers Q1 and Q4 (which also provide lower input currents), to place the minimum collector-base potentials of all the p-n-p input transistors at zero for ground-level input voltages. The lower input common-mode voltage is reached when any of these devices saturate. For the low biasing current levels that are used, this generally occurs a few tenths of a volt below ground.

"The upper limit on the input common-mode voltage of this stage is the supply voltage V_{cc} minus the sum of the base-emitter voltages of Q1 and Q2 and the saturation voltage of the current source I1 (which is a few tenths of a volt). This is the limit on the input common-mode voltage for which a small differential input voltage will be amplified. The maximum common-mode voltage that can be applied to the inputs without damage to the stage is limited by either the base-to-substrate breakdown or the reverse breakdown of the base-emitter junctions of the p-n-p transistors. The maximum differential voltage (without damage) is also limited by the reverse breakdown of the base-emitter junctions of the p-n-p transistors. These breakdowns and therefore the limits on the inputs are on the order of 90 V. In addition, the input voltages can exceed the power supply voltage. The common-mode capabilities of this circuit have been achieved at the expense of frequency response, which due to the use of p-n-p transistors, is relatively poor in comparison to that of a conventional n-p-n input stage" (With permission from the IEEE).

For the many comparators that have bipolar transistors as the input stage, note that when the input signal voltage crosses the threshold setting, one of the two input transistors switches from "off" to "on," and the other input transistor switches from "on" to "off," or the reverse. This change in conduction means that the base current also has to change correspondingly. That functioning is best served if the sources of the signal and reference voltages have a fairly low impedance at high frequencies, which may be supported by a small capacitor with a low inductance path to ground at both inputs. This can improve the delay time of the comparator and improve the rise time of its output transitions. While such a shunt capacitance may not be desirable for the signal path, it should be consistently applied to the reference input.

Of course, the input signal source impedance is of much less concern with any of the comparators having a complementary metal-oxide-semiconductor (CMOS) input stage, like the TLV7211 with its typical bias current of 40 femtoamps.

1.4   Nonideal Parameters

The functioning of an actual comparator realized in hardware brings with it a number of behaviors that are less than desirable, but always present, and maybe more or less critical to the application.

1.4.1   Input Offset Voltage

The input offset voltage is defined as the input voltage that must be applied to take the comparator to its switching threshold. Ideally this would be zero and is generally in the millivolt or less region.

1.4.2   Input Bias Current

Input bias current is defined as the current into an input port when the common-mode input voltage is zero. If the two inputs of a comparator are presented with differing source impedances, then this bias current creates an additional offset voltage.

1.4.3   Input Offset Current

Input offset current is defined as the difference in bias currents between the two inputs. This current will create additional offset voltage even when the two inputs are presented with identical source impedances.

Input offset voltage, input bias current, and input offset current are best dealt with using the same concepts as are applied to operational amplifiers having these non-ideal behaviors.

1.4.4   Input Noise Voltage

Input noise voltage is defined as the varying component of the input offset voltage. Although this is a fundamentally important parameter, it is almost never specified, probably due to the difficulty in measuring it. Measurement of this parameter may be considered as arranging to input a sine wave with a specified amplitude and very low phase noise, then measuring the phase noise of the comparator output. Details of a measurement technique based on these techniques are presented later.

1.4.5   Input Noise Current

Input noise current is defined as the varying component of input bias current. The source impedance presented to the comparator will convert this current into additional input noise voltage.

1.4.6   Oscillatory Transitions

Even an ideal comparator will respond to the even slight noise present on most real input signals by making rapid multiple transitions between one and zero as the input signal crosses the threshold voltage. The input slew rate may be increased until the multiple transitions no longer occur. The circuit layout may contribute to this oscillatory behavior by allowing the output to influence the input, and this may be corrected. The most widely used technique to avoid these multiple transitions is the addition of hysteresis, discussed in detail later.

1.4.7   Propagation Delay

Propagation delay is defined as the time interval between a change in input level sufficient to cause the output to change, and the change in the output.

1.4.8 Propagation Delay Dispersion

Changes in delay versus input level—propagation delay dispersion, can incur amplitude modulation-to-phase modulation (AM-to-PM) conversion and is covered in detail later under Non-Ideal Behavior, AM-to-PM Conversion.

References

Trump, Bruce (2012) Op Amps used as Comparators—is it okay? http://e2e.ti.com/blogs_/archives/b/thesignal/archive/2012/03/14/op-amps-used-as-comparators-is-it-okay also published in a compendium of blog posts, on page 35: http://www.ti.com/lit/sg/slyt701/slyt701.pdf

Russell, Ronald W. and Frederiksen, Thomas M. (1972) IEEE J. Solid-State Circuits, vol. SC-7, pp. 446–454, Dec. 1972

Horowitz & Hill (2015) The Art of Electronics, Third ed. Cambridge 2015 p 236–238, 809–817

Wikipedia (2020) https://en.wikipedia.org/wiki/Comparator. Accessed 2020 Jun 23

Malmstadt, Enke & Crouch (1981) Electronics and Instrumentation for Scientists, The Benjamin/Cummings Publishing Company, Inc., 1981, ISBN 0-8053-6917-1, Chapter 5.

Chapter 2
Major Applications

2.1 Major Applications of Comparators

Several of the most basic applications are mentioned in this section. There are more elaborate uses of comparators discussed in Chapter 3 **Enhancements**.

2.1.1 Digitizing an Analog Signal

This is the fundamental function of a comparator, and beyond serving as a one-bit analog-to-digital-converter (ADC), a comparator is critically important in the functioning of multi-bit converters such as a successive approximation converter, a ramp-compare converter, a Wilkinson ADC, an integrating ADC, and all types of delta-sigma converters. A direct-conversion ADC or flash converter has a bank of comparators receiving the input signal in parallel, and each comparator deals with a narrow range of the input signal.

2.1.2 Finding the Center of the Eye

When a bitstream is displayed on an oscilloscope with one or a few bit periods across the screen, the optimal sampling times and decision levels are seen as the center of the "eye," the opening around the midpoint of the bit time. The purpose of a data slicer is to make a hard decision when the signal level changes so that the slicer output represents the signal state at the center of the eye. An adaptive slicer can average the entire waveform and set its threshold at the midpoint. A further enhancement of this approach is for the adaptive circuit to capture the positive peaks

© The Author(s), under exclusive license to Springer Nature Switzerland AG 2022
M. C. Fischer, *Comparators*, https://doi.org/10.1007/978-3-030-95742-1_2

and the negative peaks and take the mean of the two to set the threshold level. The details of these enhancements are discussed later.

2.1.3   Timing Signals

When a precise timing signal is generated by a quartz crystal oscillator or other stable references, the resulting waveform is usually sinusoidal, or approximately so, and must be converted to logic levels with transition edges that can serve as timing marks. Timing signals are also recovered from rotating storage media where the amplitude may vary by more than an order of magnitude between the modes of reading and writing. These and many more analog signals must have their timing information extracted and converted to digital logic transitions.

The comparator can also be the active element in creating a timing signal as an oscillator. This is discussed under Enhancements; Oscillator.

2.1.4   FM Reception

When an angle-modulated signal like frequency modulation (FM), phase modulation (PM), or phase shift keying (PSK) is being received, after the front-end processes of down-conversion, filtering, and IF amplification, the amplitude variations on the signal that carry noise but none of the desired information must be removed. This stripping away of amplitude variations and noise components is best accomplished with a hard limiter, a function well served by a comparator. This is a critical component of every FM receiver, and many others as well.

In the FM receiver signal chain the term "limiter" has historically been used and comes from the viewpoint that the limiter stage is a more or less linear gain stage for the smallest signal levels, but is designed to clip both the positive and negative peaks of the waveform about equally as the signal strength increases. This clipping, or limiting, action tends to remove amplitude variations and makes the subsequent demodulation stage less responsive to amplitude modulation and amplitude variations caused by noise.

2.1.5   Audio Amplification

When high-power and/or high-efficiency audio amplification is needed, the design usually goes to a switching or pulse width modulation (PWM) or Class D approach. This is used in devices ranging from kilowatt-level professional audio amplifiers to cell phones and hearing aids, where energy efficiency is paramount. The fundamental element in an early stage of such an amplifier is a comparator. One approach applies a triangular waveform to the reference port of the comparator, and the analog signal to be amplified to the other port. This results in an output stream of pulses whose width varies to carry the analog information. These pulses are applied to a power output stage that switches on and off to create the high level signal with high efficiency. This is followed by a low-pass filter which attenuates the high-frequency components and passes the amplified audio to the load device, typically a loudspeaker.

The comparator speed may not be a demanding characteristic, since the switching rate is generally in the mid-megahertz or lower.

However, in such an application, the noise contribution of the comparator can be critical, as often the dynamic range of the input signals to be handled is at least 60 decibels (dB), and can range to 120 dB. At the input of the comparator, the maximum signal level might be in the neighborhood of 1.0 Vrms, which is 2.83 Vpp for a sine wave, or well over 3 Vpp for a more complex audio waveform. Starting with the 1.0 Vrms figure, then 120 dB lower is 1.0 µVrms. A white noise level of 1.0 µVrms over a 10 kHz bandwidth is 10 nV/rtHz noise density. For opamps, this is not a particularly low noise performance, the best being below 1.0 nV/rtHz, but since comparators do not have input noise specifications, this could be a challenging situation.

When the input noise of a comparator needs to be measured, a technique for doing so is described later in the section Input Noise Measurement.

2.1.6   Instrumentation

The input signal to a frequency counter can be any of a huge range of signal types, and the user of the counter expects the counter to find the fundamental or largest amplitude component and measure its frequency. That requires a signal processing element that is basically a comparator to determine the major excursions of the signal and convert them into a countable stream of digital bits. Counters may also measure period, time interval, pulse width, duty cycle, and other parameters. For each of these measurements, it is the comparator that makes the critical decision as to exactly when the next cycle occurs, when the pulse ends or when the next pulse occurs, and so on.

A similarly challenging need is the trigger circuit in an oscilloscope, which utilizes a comparator function. In the oscilloscope, the user is allowed to adjust the threshold level to select the point on the waveform where the trigger event occurs.

In both these classes of instruments, the user is given a choice of whether the comparator is DC-coupled or AC coupled. AC coupling may be considered a high-pass filter that has a corner frequency of typically a few tens of hertz, and no dc response. This is discussed at greater length under Enhancements; Decision Level Setting—Adaptive Thresholds; Average Tracking.

2.1.7   Time-Based Signal Processing

Until relatively recently, digital signal processing (DSP) has relied on converting an analog quantity into a binary word having a length of at least eight bits. This conversion is done by sampling the signal at regular intervals and converting each sample with an analog-to-digital converter (ADC). Each of these converted samples is a binary word that represents the amplitude of the analog signal at the time of the sample. The series of successive samples are then subjected to various computational algorithms such as averaging, filtering, spectral analysis by discrete Fourier transform (DFT) or fast Fourier transform (FFT).

In contrast to this, time-based signal processing compares the analog quantity to a regularly occurring ramp and effectively samples the signal when it equals the ramp value. The analog quantity is then represented by the time interval from the start of the ramp until the time when equality is reached. This time interval is only one binary bit, or pulse, and its duration carries the information to be processed. The comparison that creates this digital representation of the signal is accomplished by a comparator, thus making the characteristics of the comparator critically important to the functioning of the system.

One of the most striking advantages of time-based signal processing is that the digital hardware required is the same technology as in DSP, but the signal path is only one bit wide, instead of many bits wide. But this narrow signal path is implemented with the same circuit technology and production processes that are so well established in the microprocessor world and hence is eminently producible and very economical.

This relatively new field of application of digital logic elements has been developed to accomplish integration, differentiation, data conversion, filtering and control.

Since it utilizes pulse width modulation, it relies on comparators to convert an analog waveform to a series of digital level pulses whose widths or time intervals carry the information.

An early version of this was the delta-sigma modulator, and this approach is being researched and extended to seemingly unlimited new functions and fields of application. The 555 timer IC and its variants find some application in these systems and are explored in more detail later under Enhancements.

In time-based signal processing, the comparator is the critical element that defines the conversion of an analog signal into a digital representation. For example, if the timing is quantized as a 1.0 GHz clock rate, and a 10 bit precision representation of the analog input is desired, then the ramp that is compared to the input must have a duration of 1.0 μs. That is then a 1.0 MHz sampling rate for the analog input which will be quantized to one of 1000 levels; near the 1024 levels of 10 bit precision.

2.1.8   Human Interfaces

The human interface with digital circuitry is quite often a switch closure of some sort, as in a keyboard or push button. These switches invariably display an erratic multiple closure lasting a few milliseconds, when only a single closure was intended. This phenomenon is known as switch bounce, and must be accommodated by a circuit that can ignore the extra pulses. A comparator is very useful in implementing these circuits. The details are addressed later in the section "Switch De-Bouncers".

2.1.9   Software Implementation

Many of the concerns dealt with in this document using hardware can also be addressed in software algorithms, if the slicer is replaced with an analog-to-digital converter (ADC) that samples at many times the maximum expected rate of changes in the signal to be decoded. This approach can be cost-effective if the ADC is already present for other functions, otherwise the hardware comparator can serve well.

References

Hanumolu, Prof. Pavan (2017) University of Illinois, Urbana-Champaign, Applications of Time-based Circuits in Data Conversion, Filtering, and Control, Presented to the IEEE Solid State Circuits Society, Santa Clara Valley Chapter, 2017 Aug 17

Naraghi, Shahrzad (2009) Time-based Analog to Digital Converters", 2009. https://deepblue.lib.umich.edu/bitstream/handle/2027.42/64787/naraghi_1.pdf?sequence=1 Accessed 2019 Nov 18

Henzler, S. (2010a) Time to Digital Converters 2010, XII, 124 p., Hardcover, ISBN: 978-90-481-8627-3. http://www.springer.com/978-90-481-8627-3

Henzler, S. (2010b) Time-to-Digital Converters, Springer Series in Advanced Microelectronics 29, DOI https://doi.org/10.1007/978-90-481-8628-02, c Springer Science+Business Media B.V. 2010

Mhaidat, Khaldoon (2006) Representations and Circuits for Time Based Computation, 2006 March, Oregon Health & Science University, Portland https://scholararchive.ohsu.edu/concern/etds/9s1616164?locale=en Accessed 2021 Oct 24

Chapter 3
Enhancements

3.1 Enhancements

The circuitry surrounding a comparator can be arranged to enhance its operation and deal with the nonideal parameters.

3.1.1 Offset Nulling

The effects of all the fixed input offset parameters can be compensated with a single offset null adjustment. If this adjustment is manual, then it will suffer drift, mainly over temperature changes, but also, to a lesser degree, over long periods, days to years. In a complex system, a calibration routine and automated nulling can be implemented. An approach to automatic nulling is to remove the main input signal from the comparator, and instead feed it a low-level sine wave that is capacitor coupled, or otherwise has no offset. Then the comparator output is monitored as to its duty cycle, and the offset nulling is adjusted to bring the duty cycle to 50%.

3.1.2 Hysteresis

When oscillatory transitions occur with a comparator, one of the techniques to reduce these effects is the addition of hysteresis. This is generally a problem when the input to a slicer is changing slowly.

With low amplitude, low-frequency or slowly changing signals input to a slicer, the slew rate of the input is an important parameter to consider. For a sine wave, the slew rate at the zero crossing is:

© The Author(s), under exclusive license to Springer Nature Switzerland AG 2022
M. C. Fischer, *Comparators*, https://doi.org/10.1007/978-3-030-95742-1_3

$$\frac{dV}{dt} = V_{pp}\pi f_0 \qquad\qquad (3.1)$$

where:

$\dfrac{dV}{dt}$ is the slew rate in V/s

V_{pp} is the amplitude of the input signal in volts peak-to-peak
π is 3.14159
f_0 is the frequency of the sine wave in Hz

As a very rough rule of thumb, the slew rate of the input should be greater than 0.1 V/t_{pd}, where t_{pd} is the response time of the comparator, usually called propagation delay, defined as the time interval between the application of an input step function and the time when the output logic transition occurs. If the input slew rate is less than this, oscillatory transitions may be likely. This is due to the large amplitude output signal coupling back to the input, a situation that can be quite difficult to avoid.

When an active element functions as a comparator, its operation can have hysteresis added by connecting a positive feedback path from the output to the non-inverting input. Attenuation in this feedback path adjusts the amount of hysteresis. The term Schmitt trigger refers specifically to a comparator with hysteresis. As shown in Fig. 3.1, the addition of two resistors in a positive feedback configuration will give the stage hysteresis. That is, the input must move beyond the reference voltage by an amount that is (R1/(R1+R2)) times the output voltage before the output will change state. Then once the state changes, the input must move back that same amount to the other side of the reference voltage before the state will change back.

An inverting comparator with hysteresis is realized by connecting the input signal to the inverting input of the comparator and connecting the reference voltage to the left end of R1. Combining hysteresis with other enhancements will be dealt with along with the descriptions of those enhancements.

Representative values for the resistors are R1 = 100 Ω and R2 = 10 kΩ, which will give 1% hysteresis, as long as the source impedance driving R1 is much less than 100 Ω. This 1% figure means that the hysteresis is 1% of the output logic swing, which might likely be many percent of a smaller input signal. Should it be inconvenient to supply the input signal to R1 from such a low impedance source, an alternative connection would be to apply the reference voltage to R1 (which might

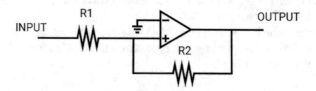

Fig. 3.1 Non-inverting comparator with hysteresis

be ground) and apply the incoming signal to the inverting input of the comparator, thus making this an inverting comparator with hysteresis.

Reference: Trump (2013)

Note that positive feedback for hysteresis is usually applied broadband but maybe high-pass filtered or AC coupled by inserting a capacitor in series with R2. This can diminish the effect of hysteresis on the threshold level, speed up the transition with an input signal of low slew rate, and eliminate oscillatory transitions. AC-coupled hysteresis allows a threshold to be sensed and held with a large amount of hysteresis, then before the next threshold crossing occurs, the hysteresis feedback is reduced as the series capacitor charges. This means that both rising and falling edges of the input signal can be sensed with large hysteresis, while the threshold for the rising edge is substantially the same as for the falling edge.

Offset Effect of Hysteresis

The application of hysteresis to a comparator stage incurs an intentional offset voltage at the input of the stage. Since the hysteresis is an attenuated version of the two logic levels of the output, then the offset may be considered to be the mid-point of the two logic levels versus the reference level, that difference attenuated by the hysteresis divider ratio. That offset voltage is:

$$V_{osh} = \frac{R1}{(R1+R2)} \cdot \left(\left(\frac{V_{oh} + V_{ol}}{2} \right) - V_{ref} \right) \tag{3.2}$$

where:

V_{osh} = Input offset voltage due to hysteresis

R1, R2 = Hysteresis feedback attenuation resistors

V_{oh}, V_{ol} = Output logic levels, high and low

V_{ref} = Reference voltage at the input of the comparator

As is clear from inspection of this relationship, the offset effect is minimized by making the reference voltage equal to the mid-point of the two logic levels. For CMOS outputs this can be accomplished by making V_{ref} equal to half the logic supply voltage; a typical situation in single-supply arrangements.

For ECL the outputs of a pair of dummy gates, one high and one low, maybe averaged to create V_{ref}. For instance, some packages in the MECL 10H series have at least one gate with differential outputs. This is true for part numbers MC10H101-105. In these cases, the inputs of a single gate may be strapped to hold the outputs constant, and the differential outputs supply V_{oh} and V_{ol}. Of course, the other gates in these packages are unaffected and may be used as needed. Keep in mind that the outputs of the 10H family, and others, are an open emitter structure, and accordingly, a pull-down resistor to a lower voltage with a current in the milli-amp range must be supplied.

Comparators with Built-In Hysteresis

Examples of comparators having built-in hysteresis are the LT1720 with specified 2–5 mV, typical 3.5 mV, TLV3501 with typical 6 mV, and LMV7219 with 7 mV, which cannot be disabled. Other comparators may also have this characteristic, so it is well to be watchful for it. If even larger hysteresis is desired, then the positive feedback discussed earlier may be added to this class of comparators. There is further discussion of the hysteresis of logic elements later in the section "Logic Elements' Thresholds and Hysteresis", as well as an example of how to add hysteresis to logic elements that have none.

How Much Hysteresis?

Choosing the amount of hysteresis to design into a slicer requires consideration of a number of points:

1. The amount of hysteresis is set by the output logic swing voltage attenuated by the voltage divider R1/(R1+R2). This ratio is often expressed as a percentage and typically ranges from 1% to 10%.
2. It would be desirable to make the hysteresis at least as large as the largest peak-to-peak noise that is expected on the input signal.
3. Hysteresis needs to be kept small because a changing signal amplitude will incur AM-to-PM due to the effect of hysteresis.
4. Even a tiny amount of hysteresis (e.g. 1%) can eliminate oscillatory transitions on small or low slew rate signals.

3.1.3 Decision-Level Setting: Adaptive Thresholds

The simplest choice for an adaptive decision level assumes that it is desired to have the decision level at the mean of the signal, the average DC level. However, depending on the types of noise and interference included, it may be more desirable to set the decision level at some point between the maximum and the minimum excursions of the signal, generally the midpoint. Depending on the amplitude statistics of the signal, this may be quite different from the mean.

Average Tracking

Given that an analog signal contains information that we want to digitize with a comparator, the next question is how to set the decision level near the mean of the signal excursions. In other words, we recognize the need to avoid influences from the DC level of the signal, which may change for various reasons and, while carrying no information, would degrade the performance of the comparator. This immediately suggests a low-pass filter, such as a series resistor followed by a shunt capacitor to find the average DC level. While the analog signal is applied to one port of the comparator, the low-pass filtered DC average of the signal is applied to the other port of the comparator as the reference level and sets the decision level. See Fig. 3.2.

For input signals that have offset voltages or low frequency wander of their average level, this circuit sets the threshold of the comparison to the longer term average of the input signal. This causes the data stream out of the comparator to represent swings above and below the average level of the input signal. In Fig. 3.2, the averaging time applied is the time constant of the R3, C1 combination, equal to R3*C1 seconds.

Figure 3.2 shows an inverting connection with the average-tracking enhancement. A non-inverting version is realized by exchanging the two inputs of the comparator so that the input signal is applied directly to the non-inverting input of the comparator, and the junction of R3 with C1 is connected to the inverting input of the comparator. Note that this junction of R3 with C1 is the node that develops the reference threshold voltage for the comparator. While the sense of the comparison is inverting, the decision level is at the mean of the incoming waveform.

These arrangements, whether inverting or not, burden the comparator with the full range of dc levels of the incoming signal, seen by the comparator as a common-mode voltage. Since all comparators have some sensitivity to changes in common-mode voltage, usually expressed as causing changes in the offset voltage and/or propagation delay of the comparator, this arrangement would be well to avoid.

The more desirable alternative, which can be shown to have identical performance while avoiding the common mode burden, is the DC block, or high-pass structure, also called, "ac coupling". The simplest form of this is a series capacitor and a shunt resistor. The shunt resistor should be returned to the same voltage as the reference voltage applied to the other input port of the comparator, generally the midpoint of the supply. Clearly, this reference voltage becomes the common-mode voltage for the comparator. See Fig. 3.3.

The transfer function of this non-inverting circuit responds identically to a connection where R3 and C1 are swapped in Fig. 3.2, the input is applied to C1 and the (−) input of the comparator is grounded, that is, C1 in series and R3 to ground. This

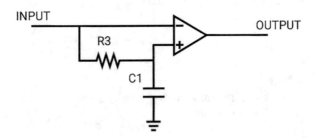

Fig. 3.2 Average-tracking threshold, inverting comparator

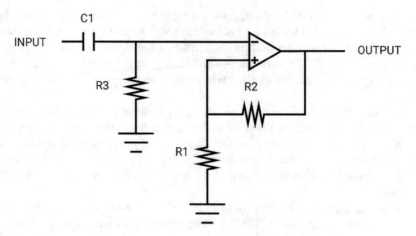

Fig. 3.3 Average-tracking inverting comparator with hysteresis

has the advantage of not subjecting the comparator to the common-mode vagaries of the signal average voltage.

Low-pass filtering to create the threshold voltage versus high-pass ac coupling is compared in a simple circuit analysis below that shows they are identical with respect to the differential signal seen at the comparator inputs.

First, the high-pass AC coupling, consisting of a series capacitor followed by a shunt resistor, can be shown to have a transfer function of:

$$V_1 = V_i \frac{sRC}{sRC+1} \tag{3.3}$$

where:
V_1 is the differential voltage seen by the comparator inputs with ac coupling,
V_i is the input voltage to the stage,
R and C are the component values.

With the low-pass filter, consisting of a series resistor followed by a shunt capacitor applied to the input signal and used to set the threshold, and having the same value resistor and capacitor as the high-pass filter, the comparator sees a differential signal with a transfer function of:

$$V_2 = V_i - V_i \left(\frac{1}{sRC+1} \right)$$

where:
V_2 is the differential voltage seen by the comparator inputs with the low-pass filter,
V_i, R, and C as above.

With manipulation:

$$V_2 = V_i \left(\frac{sRC + 1 - 1}{sRC + 1} \right)$$

And then:

$$V_2 = V_i \left(\frac{sRC}{sRC + 1} \right) \tag{3.4}$$

which is equal to V_1.

These two arrangements differ in that the high-pass coupling keeps the comparator from seeing any changes in common-mode input, a desirable feature. The AC coupling is likely to be seen as resulting in a data pattern-dependent threshold. While this is true, it is equally true of the low-pass filter, simply because their operation is identical for the same value resistor and capacitor.

This arrangement is combined with positive feedback for hysteresis in Fig 3.3.

This circuit configuration is a highly desirable choice, as long as it meets the system requirements, and this circuit avoids the hysteresis offset on the threshold voltage. The hysteresis effect as an offset voltage was quantified and discussed in more detail in an earlier section.

Whenever hysteresis is applied to a comparator, the actual decision level splits into two different levels, one for a rising input, the other for a falling input. The difference between these two levels is the amount of hysteresis. Typically one of these two levels will be essentially unchanged from the condition without hysteresis. Then the other decision level will be displaced from the reference voltage by the amount of the hysteresis.

The non-inverting average-tracking comparator with hysteresis in Fig. 3.3 tends to keep the decision levels spaced about equally above and below the non-hysteresis reference level.

Figure 3.4 shows the simplest average-tracking threshold non-inverting comparator with hysteresis, but this simplicity incurs a significant interplay between the source impedance of the driving circuit and the amount of hysteresis. Ideally, this topology should be driven from a high impedance so that R1/R2 sets the hysteresis, while the current drive times R1 sets the voltage drive to the stage. This source impedance needs to have a finite, controlled value to operate with C1 to set the high-pass corner frequency or alternatively the time constant of the average tracking function. This high-drive impedance has the further benefit of making the required value of C1 smaller. Therefore, if the high drive impedance can be conveniently accommodated, this is a clever implementation, and this circuit avoids the hysteresis offset on the threshold voltage.

Figure 3.5 shows an average-tracking non-inverting comparator with hysteresis that separates the determination of the averaging time constant from the setting of hysteresis. This topology is best driven from a low-impedance source, as is more

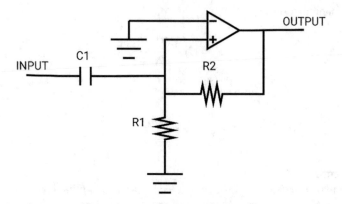

Fig. 3.4 Average-tracking non-inverting comparator with hysteresis - 1

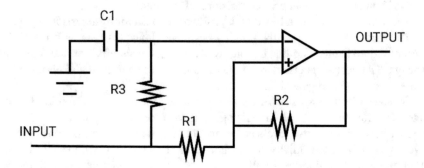

Fig. 3.5 Average-tracking non-inverting comparator with hysteresis—2

conventional practice. The less desirable aspect of this topology is that it subjects the comparator common-mode input to the full range of input signal average voltage, as does the topology of Fig. 3.2 above. However, this circuit avoids the hysteresis offset on the threshold voltage. If the effects of the common-mode variations can be shown to be acceptable, then this circuit is a desirable one for general application.

The attributes of the preceding circuits are compared in Table 3.1 to facilitate choosing the most appropriate one for a given application. Figures 3.3 and 3.4 have broad applicability and attractive functionality:

In Table 3.1, "Power" refers to whether the circuit requires dual supplies (positive and negative voltages), or maybe operated with a single supply bus. As is indicated, the circuits that require dual supplies may also be operated with a single supply bus as long as a one-half supply voltage bus is made available.

The row "Invert" shows whether each circuit is configured to be inverting, non-inverting, or may be connected to operate as either inverting or non-inverting.

Table 3.1 Attribute matrix

Figure	1.1	1.3	1.4	3.1	3.2	3.3	3.4	3.5
Power	Dual	Dual	Single	Dual[a]	Either	Dual[a]	Dual[a]	Either
Invert	No	No	Either	No	Yes	Yes	No	No
Couple	DC	DC	DC	DC	DC	AC	AC	DC
Hyster	No	No	No	Yes	No	Yes	Yes	Yes
Adapt	No	No	No	No	Yes	Yes	Yes	Yes

[a]Single with half-supply bus

The row "Couple" refers to the input path and shows "DC" for the circuits that pass the DC component of the input signal to the comparator input. In Figs. 3.2 and 3.5, the DC component of the input voltage also appears as a common-mode voltage, thereby tasking the comparator with tolerating the changes in DC level.

The row "Hyster" shows whether the circuit includes the positive feedback path necessary to supply hysteresis.

The row "Adapt" shows whether the circuit includes the adaptive threshold that tracks the average of the input signal.

Peak Mean Tracking

All the preceding adaptive tracking thresholds set the threshold level at the average of the incoming signal. While this may suffice for a simple, predictable signal, it may be worth the added complexity to determine the negative and positive peak excursions of the signal and set the slicing threshold at the voltage mid-way between these peak levels. Adding circuitry to detect both the positive and the negative peaks adds significant complexity to the slicer stage, as can be seen in the examples shown in the following figures. Advantages of this arrangement are greatly reduced dependency of the reference level on changes in data patterns, and a reduction in the effect of comparator offset voltages.

A way of setting the decision level that is adaptive to changes in the signal amplitude is to do peak detection on the positive peaks and peak detection on the negative peaks. Then a resistive divider sets the decision level at the mid-point between these two peaks, as shown in Fig. 3.6. Another case where this is desirable is when the input signal is a sequence of narrow pulses, either negative or positive. In this case, the best setting for the reference level is halfway between the positive and negative excursions of the signal, rather than the average level of the waveform.

In the case where the signal-to-noise level of the input signal to the comparator is negative, then the mean of the peaks will be determined by the noise peaks, not the signal peaks. This mean voltage then becomes the reference voltage for the comparator, and the decisions are made mostly on the noise waveform, with some influence by the desired signal. If the comparator is followed by a narrow-band phase-locked loop under these conditions, the loop will have a tendency to track the desired signal, in a noise-perturbed fashion, with a loop bandwidth that is narrower

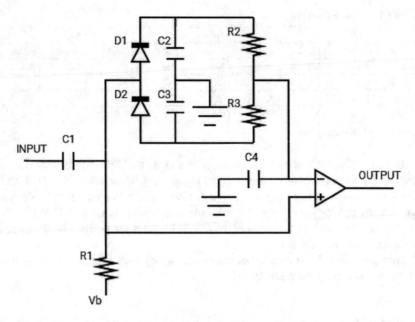

Fig. 3.6 A comparator with peak mean tracking reference, non-inverting

than when the input signal-to-noise ratio is positive. This is because the noise effectively reduces the transfer coefficient of the phase detector.

When a negative signal-to-noise ratio at the input to the comparator is the usual or most critical condition, it may be advisable to use the average tracking adaptive threshold discussed in the preceding section. This decision should be guided by a knowledge of the characteristics of the noise; that is, is it additive white Gaussian, or atmospheric impulse noise from lightning, or interfering signals with more discrete spectral components.

As mentioned in the discussion following Fig. 3.5, changes in the common-mode voltage seen by the comparator under different signal conditions are best avoided. Common-mode changes can be eliminated by applying for a dc blocking high pass ahead of the peak detectors and comparator. Here the shunt resistor is returned to the supply voltage mid-point. The sequence of operations is to high-pass, apply this signal to both one input of the comparator and to the peak detectors, then the peak detected mid-point is applied to the other input of the comparator as the reference level. See Fig. 3.6. To extend the dynamic range of such a circuit into lower levels, Schottky diodes are desirable and available as a packaged pair as a BAT54S.

The mean of the positive and negative peaks will be less dependent on data content than is the low-pass filtered reference level. It is still somewhat dependent because of the finite time constant (decay time) of the peak detection circuits. These time constants must be set short enough to allow the peak estimates to follow the anticipated rate of change in the signal amplitude. But to minimize data dependency, the time constants should be as long as allowable.

In Fig. 3.6 the high-pass function is accomplished by C1 with R1. The voltage Vb applied to R1 sets the common-mode voltage to both inputs of the comparator, and should be ground for dual supplies, or half the supply voltage for a single supply. The drive for this stage should be low impedance because the diodes D1 and D2 only conduct for a small fraction of a cycle of the input frequency, and the source must supply the charging current for C2 and C3 during this short interval. The diodes D1 and D2 catch the peaks of the incoming signal and store the positive peak voltage in C2 and the negative peak voltage in C3. The decay time constant of these peak levels is set by the product of C2*R2 and C3*R3. If R2 and R3 are equal, then the junction of R2 and R3 has a voltage midway between the two peaks, and applies this to the comparator negative input as the reference voltage. To make this circuit function better with small signals, the diodes D1 and D2 can be Schottky barrier diodes such as BAT54S or BAR43S. The capacitor C4 needs to be located near the negative input of the comparator, where it provides a low impedance at high frequencies (much lower than the parallel combination of R2 and R3). The input stage of the comparator can run at full speed when it sees a low impedance at high frequencies at both inputs.

The bias current of the negative input of the comparator must be supplied through R2 and R3, and their values should be limited to that which can supply this current with acceptable offset voltage. That current must also flow through D1 and D2. Therefore at low signal conditions, where D1 and D2 are essentially non-conducting, the bias current can create an otherwise unexpectedly large offset voltage.

For a special case where the signal is expected to be asymmetrical, or if for some other reason it is desired to set the reference level nearer either the positive or negative peak value than the mean, that is easily accommodated by adjusting the values of the resistors, R2 and R3 to something other than equal.

The circuitry in Fig. 3.6 is perhaps the simplest way of accomplishing the peak detection function. The limitation of this simple approach is that input signal levels of less than about 1.2 V peak-to-peak will not have a significant effect on the reference voltage. There are several more complex circuits that can do a better job of peak detection over a wider range of signal amplitudes and at higher frequencies. These involve adding a transistor with each diode, or replacing each diode with two transistors, or adding an operational amplifier with a pair of diodes for each detector, as shown in the circuits to follow.

The peak detectors in Fig. 3.7 can be operated down to input signal levels of as low as 0.10 Vpk-pk and extend to higher frequencies than Fig. 3.6. Note that the resistor to Vb following C1 at the input node has been removed. It is no longer necessary, because R1 and R2 keep D1 and D2 in conduction most of the time, and Q1 and Q2 can supply the bias current to the negative input of the comparator. The positive input of the comparator is now set to the mid-point of the supplies and its bias current is supplied by this current flow through R1, D1, D2, and R2. Accordingly, the peak detection function is now occurring at the emitters of Q1 and Q2 which supply the current to charge C2 and C3, thereby reducing the load on the input signal. Further, the high-pass input function is now set by C1 working with the parallel

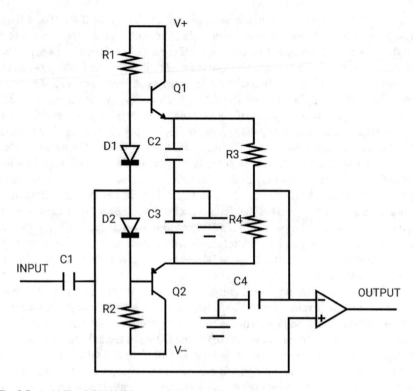

Fig. 3.7 As in Fig. 3.6 but with more elaborate peak detectors

combination of R1 and R2. The diodes D1 and D2 should be standard PN junction diodes so that their forward voltage is similar to the base-emitter drop of Q1 and Q2.

The peak holding time constant is set by C2 with R3 and C3 with R4.

A more detailed circuit example of this technique is shown on the Linear Technology LT1394 and LT1671 datasheets and the application note AN72. The circuit in the application note claims to accommodate input level variation over a 2 mV to 175 mV range with a 45 MHz input signal. Since the base-emitter junctions are used as peak detection diodes, the reverse breakdown voltage of the junction must not be exceeded by the largest signal condition. This limitation may be best met by limiting the total supply voltage to 5 V. Otherwise a signal diode of sufficient ratings must be used.

In Fig. 3.8, a resistor to a bias voltage, Rb to Vb, must be added, much as in Fig. 3.6, at the downstream end of C1. The voltage Vb applied to Rb sets the common-mode voltage to both inputs of the comparator, and should be ground for dual supplies, or half the supply voltage for a single supply. The high-pass filter function is performed by the combination of C1 with Rb, as R1 in Fig. 3.6. This resistor also provides a path for the difference in base currents of Q1 and Q2, as they will likely have differing betas.

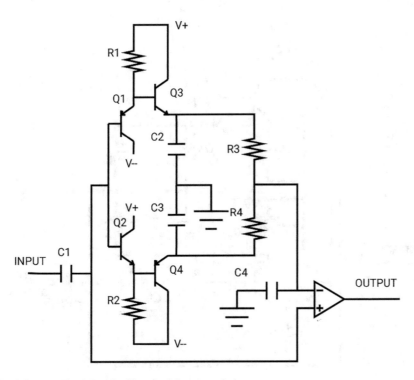

Fig. 3.8 As in Fig. 3.7 with still more elaborate peak detectors

The peak detectors in Fig. 3.8 can be operated down to input signal levels in the low millivolt peak-to-peak range, and at still higher frequencies. The pair of transistors at each polarity of peak detection, Q1 and Q3 for positive peaks and Q2 and Q4 for negative peaks, is a variant of the Darlington connection that gives the product of the current gain of the individual transistors improving the frequency response and speed of operation while further increasing the input impedance of the stage. The input impedance of this stage is now the parallel combination of R1 and R2 multiplied by the beta of Q1 and Q2.

Here as in Fig. 3.7 each base-emitter junction of Q3 and Q4 will be subjected to the full peak-to-peak value of the input waveform as a reverse voltage, thus requiring the maximum input signal to be limited to about 5 V peak-to-peak. Note that, at low signal levels, Q1 and Q2 will be conducted continuously, while the peak detection function occurs at the base-emitter junctions of Q3 and Q4, which only conduct on the peaks of the input waveform.

When operating at low signal levels, offset voltages become significant, and a way to minimize those errors is to have all four transistors, Q1–Q4, in a single chip with specified Vbe matches such as the HFA3096 (originally an RCA part, now from Intersil, part of Renesas). That chip has an absolute maximum spec for base-emitter reverse bias of 5.5 V. The circuit of Fig. 3.8 can reach this limit with an input signal of 5.5 Vpp, so the drive circuitry must be configured to limit at 5 Vpp or less

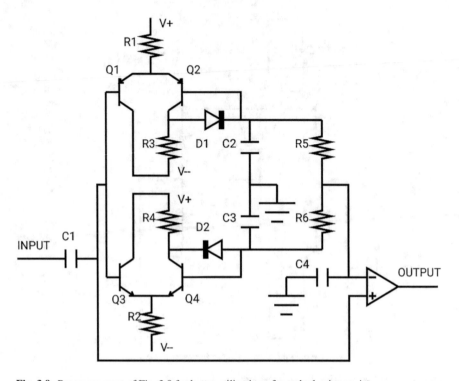

Fig. 3.9 Rearrangement of Fig. 3.8 for better utilization of matched pair transistors.

as a worst-case tolerance limit. The circuit in Fig. 3.9 operates in a quite similar fashion and must be analyzed based on the reverse Vbe specs of the transistors employed. For widely used transistors, the Vbe Max spec ranges as low as 5 V and as high as 6 V.

In Fig. 3.9, the npn and pnp pairs that have matching specs in the HFA3096 are brought together to take better advantage of the matched behavior to reduce offset voltages. This connection still has the same Vbe limits on large input signal swings as in Fig. 3.7. Another disadvantage is that there will be less current available to charge C2 and C3 than in Fig. 3.8. This can be seen from inspection of the current path for charging C2, for example. The maximum charging current will be the current flow through R1 minus the current flow through R3. Unfortunately, the current through R3 will be the highest when the largest signals are detected, and when a large charging current is most needed.

Of course, the differential pairs in Fig. 3.9 may be swapped, as shown in Fig. 3.10.

This results in a maximum charging current of the peak holding capacitors C2 and C3 that is supplied by R3 and R4 alone, which will be larger if V+ and V− are large. This maximum charging current will still be less than the quiescent current through R3 and R4.

Since all the circuit arrangements from Figs. 3.7 through 3.10 suffer the maximum peak-to-peak signal limitation of a single transistor reverse Vbe limit, we will

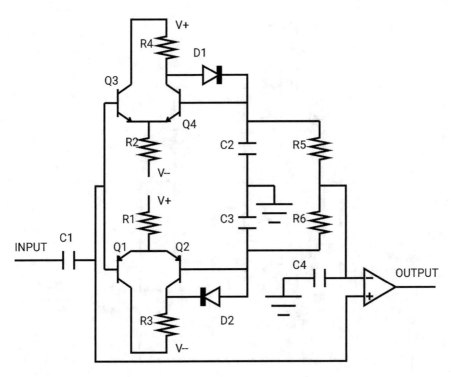

Fig. 3.10 Rearrangement of the differential pairs of Fig. 3.9

explore circuits that employ signal diodes that can have an order of magnitude higher reverse voltage limits, in the path to charge the peak holding capacitors.

To avoid the constraint of transistor reverse Vbe limit, the circuit in Fig. 3.11 replaces Q3 and Q4 in Fig. 3.8 with diodes D1 and D2 that can have a much higher reverse voltage specification, such as the BAW56 with a 70 V rating. This arrangement can be seen as quite similar to Fig. 3.7, with each diode and transistor swapped, and offering the advantage of allowing an order of magnitude larger input signal, while its other characteristics are similar.

A disadvantage of these changes is that while in Fig. 3.7, Q1 and Q2 can supply large charging currents for capacitors C2 and C3, here in Fig. 3.11, those capacitors can be charged only by the currents flowing in R1 and R2, likely smaller by at least an order of magnitude.

Once again, the peak holding time constant is set by C2 with R3 and C3 with R4. And since this stage has a fairly high input impedance, a resistor will almost certainly need to be added from the down-stream side of C1 to the mid-supply point, to work with C1 to establish the input high-pass corner frequency, as R1 in Fig. 3.6, to determine the common mode voltage for the comparator, as well as to allow for unequal betas of Q1 and Q2.

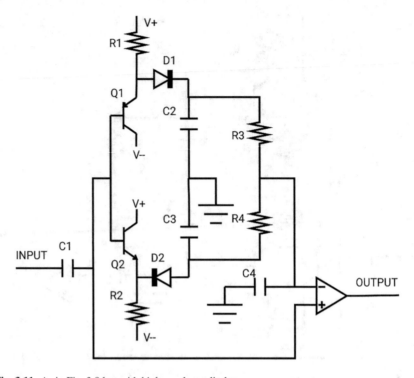

Fig. 3.11 As in Fig. 3.8 but with higher voltage diodes

Figure 3.9 suggests that instead of matched pairs of transistors, we might get all the advantages and no disadvantages if we change to a pair of operational amplifiers in place of the transistors. That is what is shown in Fig. 3.12.

Figure 3.12 shows that to use op-amps, U1 and U2, to improve the peak detection performance, it is necessary to add two more diodes, D1 and D3. The comparator remains at U3. For a positive-going signal, U1 charges C2 through D2, while D1 is back-biased and presents an open circuit, allowing R1 to close the loop around U1. Then when the input swings negative, D1 turns on and D2 turns off. The feedback path through D1 keeps the op-amp from going into an open-loop overload state that would have a lengthy recovery time. The opposite polarity effects occur with U2, D3, D4, R2, and C3. Once again, the worst case reverse voltages will be borne by the diodes and can be expected to equal the peak-to-peak value of the input signal; completely avoiding the transistor reverse Vbe limitation.

To set the peak holding time constant, it must be noted that for C2, both resistors R1 and R3 will be parallel discharge paths, and similarly, for C3 there are R2 and R4. The voltage level to which this discharging occurs is, on the average, the supply mid-point, which is also the common mode voltage for the comparator. To set this voltage and to provide for bias currents of the two opamps and the comparator, as well as to set a high-pass corner frequency with C1, the addition of a resistor Rb from the right-hand end of C1 to a bias voltage, Vb, must be added, much as in

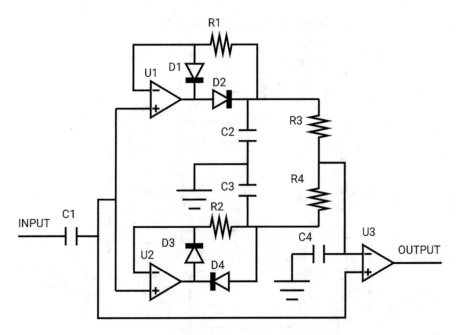

Fig. 3.12 Using op-amp peak detectors for a peak tracking adaptive threshold

Fig. 3.6. The voltage Vb applied to Rb sets the common mode voltage to both inputs of the comparator, and should be ground for dual supplies, or half the supply voltage for single supply.

For representative circuit values, R1 and R2 might be 10 kΩ and R3 and R4 might be 100 kΩ. A fifth resistor to ground (or Vb) from the down stream side of C1 is needed to set the high-pass corner frequency, provide a path for bias currents, and define the common mode voltage, as R1 in Fig. 3.6.

For implementation as an analog integrated circuit, the complexity of Fig. 3.13 offers performance similar to Fig. 3.12 without requiring operational amplifiers.

In Fig. 3.13, the diodes D3 and D4 protect the base-emitter junctions of Q3 and Q4 from damaging reverse voltage. The connection of R5 and R6 from base to emitter, rather than base to ground, carries the peak voltage stored on C2 and C3 to the diodes D3 and D4 to avoid reverse voltage on the base-emitter junctions of Q3 and Q4. The forward drops of D3 and Q3 are balanced by the drops of D1 and Q1; and similarly D4 and Q4 by D2 and Q2. The transistors Q1 and Q2 also serve to increase the input impedance and allow R3 and R4 to be lower values to supply drive to Q3 and Q4. The current gain of Q3 and Q4 provide higher drive currents to charge C2 and C3. The parallel combination of resistors R1 and R2 work with C1 to provide the high-pass corner frequency; and if they are equal value, they set the no-signal operating point and common mode voltage to the comparator at the mid-point between the two supply rails.

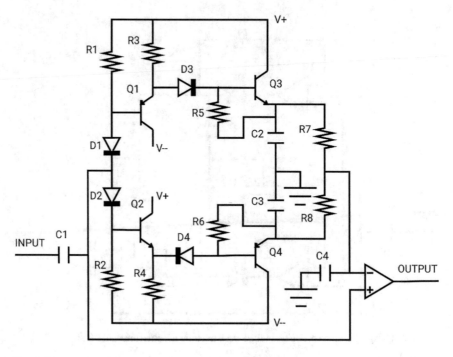

Fig. 3.13 As in Fig. 3.8 but with still more elaborate peak detectors

The peak-holding time constant is set by C2 with R7 and by C3 with R8. Although it may seem that R5 and R6 provide discharge paths, they do not, because D3 and D4 are off except during the times that signal peaks are charging C2 or C3.

This still allows R7 and R8 to be unequal if it is desired to have the reference level offset from the mean of the positive and negative peaks. Note that in order to accommodate the two diode drops in D1 and Q1, the supply voltage V+ must be at least 2.0 V greater than the positive peaks of the input signal, and correspondingly with D2 and Q2, V− must be at least 2.0 V more negative than the negative peaks of the input signal.

The base current to Q3 (and Q4) is provided by the current through R3 (and R4) during signal peaks when the charging currents for C2 (and C3) are flowing. This means that the supply voltages must be enough larger than two diode-drops above the signal peaks to supply this current flow.

For example, if the maximum signal level were 5 V peak-to-peak, the supply voltages were +5 V and −5 V, and we desired a peak charging current of 10 mA to C2 and C3, then at this peak signal instant, there is the following situation: The voltage on C2 will be 2.5 V, the voltage at the emitter of Q1 will be 2.5 + 1.3 = 3.8 V. Let's assume that Q3 has a beta of 100, then its base current will be 100 μA. With this 100 μA supplied through R3, then the voltage across R3 will be 1.2 V and its value must be set to 12 kΩ. Then during small-signal or quiescent conditions the voltage across R3 will be 3.7 V and the current flow through R3 will be 308 μA, still a reasonable drain. But this occurs in a case where the total supply voltage is twice the peak-to-peak input signal voltage.

If the complexity of Fig. 3.13 can be supported by performance requirements, this arrangement is highly desirable, even if built of discrete components, and especially if integrated. The circuit in Fig. 3.13 may be breadboarded with an HFA3096 to test its performance with transistors that share a chip. In the case where this arrangement is integrated, the only off-chip components required would be the capacitors.

In all the circuit examples of Figs. 3.7 through 3.13, the transistors are shown as bipolar, and in Figs. 3.7 and 3.8, the diode action of the base-emitter junction was necessary to hold the peak values on the capacitors. The obvious question of how CMOS circuitry can provide these functions will now be addressed. The transistors in Figs. 3.9 and 3.10 may certainly be metal-oxide semiconductors (MOS), with appropriate turn-on voltages, and the op-amps in Fig. 3.12 may also be CMOS.

In an enhancement mode MOS field-effect transistor, the term "turn-on voltage" meaning the bias applied to the gate relative to the source, which is also called "threshold voltage" with the symbol V_{th}, is the voltage at which the channel of the FET begins to conduct. This use of the term "threshold" is of course distinct from the same term applied to a comparator element, in this case describing a characteristic of an enhancement mode MOS transistor.

One of the most attractive circuits, that of Fig. 3.13, may also be implemented with CMOS transistors as is shown in Fig. 3.14.

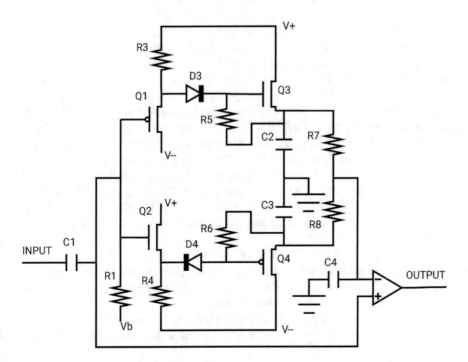

Fig. 3.14 As in Fig. 3.13 but with MOS transistors

In Fig. 3.14, the diodes and resistors at the input of Fig. 3.13 are no longer needed, but a resistor to a bias voltage, R1 to Vb, must be added, much as in Fig. 3.6. The voltage Vb applied to R1 sets the common-mode voltage to both inputs of the comparator, and should be ground for dual supplies, or half the supply voltage for single supply. The polarities of the devices are positive MOS (PMOS) for Q1 and Q4, and negative MOS (NMOS) for Q2 and Q3.

This circuit can have a very high input impedance, essentially set by R1, which with C1, also sets the input high-pass corner frequency. R3 and R4 may be of moderate value, based on the maximum frequency to be used, and the capacitances of the associated devices. R5 and R6 may be of a relatively high value, greater than R3 and R4. R7 and R8 will also be of a moderate to high value, as they work with C2 and C3 to set the peak holding time constant.

The operation of the circuit in Fig. 3.14 may be understood by considering a positive-going input excursion. This will cause Q1 to conduct less, and the voltage at its source, the node with R3 and D3, will rise. D3 will start to conduct, if the voltage on C2 is less than the incoming excursion. When D3 conducts, this causes Q3 to conduct and charge C2 until the voltage on C2 represents the incoming excursion. A negative-going excursion at the input follows a corresponding path through Q2, D4 and Q4 to charge C3. This explanation is simplified by ignoring the V_{th} of the transistors and the forward voltage of the diode as they can be arranged to offset one another to some extent.

Ideally, the voltage drops can be canceled by designing Q1 to have a V_{th} equal to the sum of the diode drop and V_{th} of Q3. Clearly this means that Q1 and Q2 should have the same V_{th} (of opposite sign). Then similarly Q3 and Q4 should also have matching V_{th} (again of opposite sign) but different from Q1 and Q2. If this is to happen in an integrated version, then the two NMOS transistors must have different process steps, and similarly for the two PMOS transistors. If it is desired to use the same process steps for all the transistors, then it would be helpful to make the diodes Schottky to minimize their forward voltage. Another alternative is to add the diodes and resistors to the input as shown in Fig. 3.13. This has two effects: first it is desirable to have the V_{th} of all four transistors to be the same, and second the input impedance of the stage becomes the parallel combination of the two added resistors.

It would be tempting to consider the CD4007 as a candidate for breadboarding the circuit of Fig. 3.14, but some of the six transistors in that device are connected internally making them inaccessible to this purpose. The input pair with pins 6, 7, and 14 can serve as the input pair in Fig. 3.14, as well as the pair with pins 2, 3, and 4. But in Fig. 3.14, the devices Q3 and Q4 require independent transistors that are not available in the CD4007.

In all the circuits of Figs. 3.6, 3.7, 3.8, 3.9, 3.10, 3.11, 3.12, 3.13, and 3.14, hysteresis may be added by connecting a large resistor from the output of the comparator to the positive input along with inserting a smaller resistor from the positive input to C1. See Fig. 3.1 and the associated text.

Without hysteresis, each of the circuits of Figs. 3.6, 3.7, 3.8, 3.9, 3.10, 3.11, 3.12, 3.13, and 3.14 may have the positive and negative inputs exchanged to make it inverting. In this case, adding hysteresis is more complicated and will likely require the addition of a voltage-follower stage to buffer the reference voltage to the positive input with its connections that implement hysteresis.

3.1.4 Ground Differences

The ac-coupled circuits of Figs. 3.3 and 3.6 also offer the first step of a path toward a design that can ignore differences in the ground potential between the source and the comparator. A typical case of this is when the comparator is desired to be operated in a single supply mode—often the situation when the output of the comparator drives logic circuits, while the input comes from analog circuitry that may have its waveform more or less centered about its ground. In cases like this, the analog ground may be at a different potential from the digital ground. This situation may sound like a potential for failure, but in fact, the comparator's balanced inputs and common mode rejection can offer a way to deal with it. Of course the single-input comparators mentioned in other sections are not applicable to these techniques.

To allow the common-mode rejection of the comparator to function to the advantage of the needs of the circuit, each of the differential inputs of the comparator must be connected to identical networks that occupy the path to the input signal. If the input signal is available in a differential form, this is ideal. Even if the input signal is single-ended or unbalanced, this technique still allows for ground potential differences to be substantially ignored. The fact that many comparators exhibit a change in delay versus a change in common-mode voltage means that changes in the ground difference potentials may be superimposed on the timing of the output of the comparator. These effects on stage delay are discussed further in Chap. 4, Non-ideal Behavior, Dispersion.

In Fig. 3.15, we see that the input signal is applied across C1 and C2, which provide high frequency paths to the comparator inputs. R1 and R2 provide a high-pass filter function operating with C1 and C2, while allowing the determination of the common-mode voltage to the comparator input. To maintain the common-mode rejection of the comparator, the values of C1 and C2 should be equal as well as the values of R1 and R2. The common-mode bias level of the comparator is supplied by the bias voltage Vcm, which should be set to a value that meets the input specifications of the comparator, considering both the Vcm dc value, plus the excursions of the incoming signal, including the worst case excursions of the ground differences.

The decision level for the circuit in Fig. 3.15 is the average of the input waveform, as was discussed in connection with Figs. 3.2 and 3.3, that is an average tracking adaptive threshold. This circuit adds no hysteresis beyond that which may be integral to the comparator. The addition of positive feedback components to increase the hysteresis incurs a tradeoff with their resulting degradation of the common-mode rejection of the stage, that makes it more sensitive to potential differences in the two grounds.

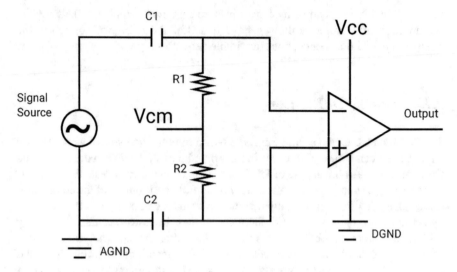

Fig. 3.15 Comparator with ground differences

3.1.5 Negative Feedback

In cases where the input is known to be a fairly symmetrical waveform, such as a sine wave, then negative feedback can be connected around the comparator stage to reduce the effects of offset voltages at the comparator input. This negative feedback path must be low-pass filtered with a cutoff frequency well below the lowest input signal frequency. Otherwise the negative feedback would operate to reduce the amplitude of the signal as seen by the differential inputs of the comparator proper. If the reference level is at the mid-point of the two output states, then the filtered negative feedback operates to take the average of the output waveform and use that average to shift the input average voltage toward a level that will tend toward a 50% duty cycle, symmetrical output. Obviously, if the desired output signal is not a 50% duty cycle, then negative feedback is not appropriate. An example of such a connection is shown in Fig. 3.16, where the low-pass function is accomplished by R3 with C1.

Combining Negative Feedback with Hysteresis

The desirability of two of the possible enhancements, negative feedback and hysteresis, makes it tempting to employ both. But doing so incurs a perhaps unexpected behavior. If a comparator stage has both hysteresis and negative feedback, this arrangement, in the absence of signal, can create an oscillator. The oscillatory condition is met whenever the negative feedback is larger than the hysteresis; or alternatively, whenever the DC attenuation in the negative feedback path is less than the

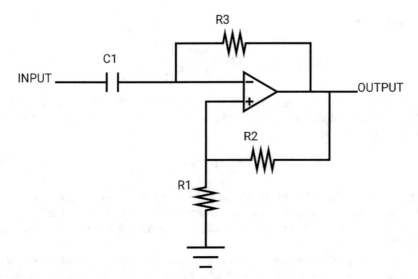

Fig. 3.16 Negative feedback, hysteresis, average tracking, inverting

attenuation in the positive feedback, or hysteresis, path. This would generally be the case, where both enhancements are present, as in Fig. 3.16, where the negative feedback has no attenuation at low frequencies, and the positive feedback is typically greatly attenuated to give a small amount of hysteresis.

A third enhancement, average tracking with high-pass or ac coupling, is easily added to a comparator that has negative feedback and hysteresis. The circuit in Fig. 3.16 offers all three of these enhancements, while being an inverting configuration. This circuit is well suited to "squaring up" a sine wave input where a range of signal levels and waveforms need to be accommodated while driving digital circuitry with standard levels.

Should the oscillatory behavior in the absence of input signal be unwanted, then the circuit in Fig. 3.3 may be preferable, as it has no negative feedback.

In the circuit of Fig. 3.16, the averaging tracking time constant is set by C1*R3, as well as the negative feedback low pass frequency or time constant.

The oscillatory behavior is not necessarily a problem; for instance, if there is always an input signal present, then the self-oscillation of the stage will never be apparent. There may be other situations where the presence of an output signal even in the absence of an input signal is either desirable or of no concern.

The frequency of the self-oscillation is set by:

V_o, the output voltage swing in V peak-to-peak.

V_h, the hysteresis band in V peak-to-peak.

R, the negative feedback low pass filter series resistor in Ω.

C, the negative feedback low pass filter shunt capacitor in farads.

τ, the period of the self-oscillation in s.

f_{so}, the frequency of the self-oscillation in Hz, $= (1/\tau)$.

in the formula:

$$f_{so} = 1/\tau = \frac{Vo}{Vh}\left(\frac{1}{4RC}\right)$$ (3.5)

For example, with a low-pass filter of 10 kΩ and 100 nF, the filter time constant is 1 ms and has a cutoff frequency of 160 Hz, and with an output swing of 4.0 V and hysteresis of 10 mV:

$$f_{so} = 100\,\mathrm{kHz}$$

The formula is derived from considering the slew rate at the low pass capacitor. During one half of the period of the self-oscillation, the voltage on the capacitor changes by an amount equal to the hysteresis band, thus having a slew rate of $V_h/(\tau/2)$.

The signal driving this slew rate is the current through the low-pass filter resistor. The voltage across this resistor is one-half of the output swing because the capacitor stores the approximate mid-point of the output swing. This makes the current through the resistor $(V_o/2)/R$. The resulting slew rate is then this current divided by the capacitance: $V_o/(2RC)$. Then these two expressions for the slew rate are equated and solved for τ, then f_{so}.

For small input signals whose peak-to-peak magnitude are less than the hysteresis band, the input signal will have a tendency to injection-lock the self-oscillation. The boundary condition for injection locking is:

$$1/\left(1-\left(V_i/V_h\right)\right) \ge \left(1/2\right)\left(\left(f_i/f_{so}\right)-\left(f_{so}/f_i\right)\right)$$ (3.6)

Where V_i is the input voltage peak-to-peak, V_h is the hysteresis band in volts peak-to-peak, f_i is the incoming signal frequency in Hz, and f_{so} is the self-oscillation frequency in Hz. If this inequality is satisfied, then injection locking will occur, in the absence of noise. This relationship is derived from the conditions:

1. If f_i approaches f_{so}, then lock will occur with V_i/V_h approaching zero.
2. If f_{in} is orders of magnitude different from f_{so} (larger or smaller), then V_i/V_h needs to approach unity to lock.

Algebraic manipulation of the above inequality can yield a more easily evaluated, if less obvious version:

$$\frac{V_i}{V_h} \ge \frac{\left(f_i - f_{so}\right)^2}{\left(f_i^2 + f_{so}^2\right)}$$ (3.7)

There are four cases that occur over the range of input signal amplitudes:

1. $(V_i/V_h) \ge 1$ Input signal fully captures the comparator, no self-oscillation.

2. (V_i/V_h)<1 but satisfies the IL criterion Self-oscillation will exist and injection locking will probably occur.
3. (V_i/V_h)<1 but IL criterion not satisfied Self-oscillation will exist and injection locking will not occur.
4. No input signal–Self-oscillation occurs.

Since the input-to-output phase shift of the comparator stage is a function of the input signal amplitude, in this case, where (V_i/V_h) <1, this is generally a situation to be avoided. That consideration then sets the dynamic range of useful operation of the comparator stage that has both negative feedback and hysteresis to be from the hysteresis band up to the maximum allowable drive level.

AC coupled hysteresis will not prevent this oscillation from occurring, but its parameters will affect the resulting frequency.

3.1.6 Oscillators

In the preceding section, the concept of combining both negative feedback and hysteresis was discussed along with the propensity for such a circuit to oscillate. When a simple oscillator is needed, the circuit of Fig. 3.16 (with the input grounded or with injection locking) constitutes an RC oscillator with modest stability performance. Stability is improved by setting the hysteresis to be large, that is, much of the common-mode range of the input. Note that the frequency may be varied over a wide range by adjusting one of the resistors that set the hysteresis. In Fig. 3.16 the resistor R3 is also a convenient component to vary for frequency control.

A quartz crystal may be used as a much more stable frequency control element with a comparator oscillator, and similarly, an L-C resonator may also be used.

The most frequency stable oscillator that can be constructed with a comparator is one in which the frequency is controlled with a quartz crystal. In Fig. 3.17, a quartz crystal, Y1, is connected as the feedback path around a comparator, U1, to control the frequency of oscillation. Since there are many variables to be managed in such a circuit, we will describe a general starting point for component values in the design. Start with R1 = 1 kΩ, and raise it to reduce the signal level at the input to the comparator to the vicinity of 1 V peak-to-peak. Y1 will be specified to operate with a certain "load capacitance", and this value is a good starting point for C1. Start with C2 at a value that gives it a capacitive reactance at the crystal frequency of 100 Ω. Should the oscillation fail to start, reduce the value of C2. The purpose of R2 is to supply DC bias to the comparator input, and to keep the DC average voltage across the crystal minimal. For CMOS this can be satisfied with a large value such as 1 MΩ. A large resistor from the output back to the inverting input ensures that the oscillator will start up when power is applied. This negative feedback sets the comparator to its crossover point at power-up, where it can function as a quasi-linear amplifier, and circuit noise will initiate oscillation. Keeping R2 large avoids having it add loading to the crystal that would lower its in-circuit Q factor.

Fig. 3.17 Quartz crystal oscillator using a comparator

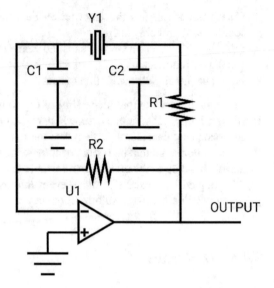

It may appear that the feedback is negative and therefore not likely to oscillate. However, with the crystal offering a small inductance, the connection with C1 and C2 will have a 180° phase shift, and this combined with the inversion in the comparator gives 360° of phase shift around the loop, as needed for oscillation.

There are numerous examples of comparators operating with quartz crystals as oscillators in various application notes from semiconductor manufacturers, but beware, they nearly all seriously load the crystal with a resistive component that will reduce its Q factor, in some cases, drastically. It is the fact that crystals have Qs of 10,000, 100,000 or even 1,000,000 that is the reason we go to the expense of using them, so let us not degrade them with poor circuit design.

Another configuration of a comparator circuit as an oscillator can employ an LC resonant circuit to control the frequency. This will be less stable than the quartz version, but more stable than the RC oscillator in Fig. 3.16.

For the circuit in Fig. 3.18, component values may be chosen starting with R1 = 1 kΩ, and the reactances of the inductor and capacitors at the desired frequency: 1 kΩ for L1, 900 Ω for C1 and 100 Ω for C2. For fine-tuning of the frequency, vary C1. As in the quartz oscillator, raise R1 to bring the signal level down to about 1 Vpp at the input to U1, and if oscillation fails to start, reduce C2. The large resistor R1 in Fig. 3.17 is missing from Fig. 3.18 because its function is performed by the DC path through R1 and L1, where in Fig. 3.17 the crystal blocked this DC path.

Any of these oscillators may be made to have voltage control of frequency or VCO, a voltage-controlled oscillator. This is done by making the C1 component in Figs. 3.16, 3.17, 3.18 and 3.19 variable. This can be done by using a voltage variable capacitance diode or varactor for C1, or better, by making C1 the series combination of a fixed capacitor and a varactor.

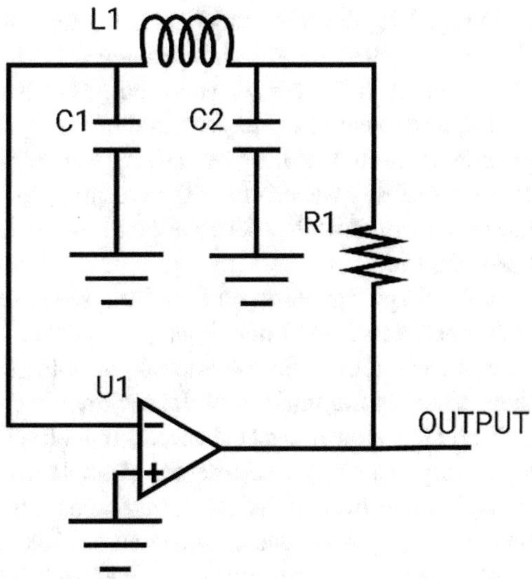

Fig. 3.18 LC oscillator
using a comparator

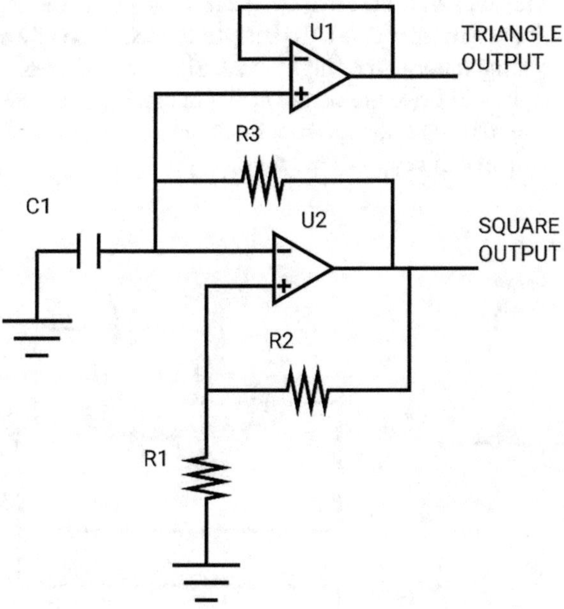

Fig. 3.19 RC oscillator with triangle output

The oscillators in Figs. 3.16, 3.17 and 3.18 have only a square wave (digital) output, but a triangle wave (analog) output can also be easily obtained. Figure 3.19 shows both triangle and square outputs.

In Fig. 3.19, the component U2 is a comparator with hysteresis set by R1 and R2. That comparator charges and discharges C1 as its hysteresis takes it between the two states of the square wave output. This charging and discharging of C1 is, of course, a segment of the exponential charging curve of capacitor voltage as the currents flow through R3, driven by the square output. If the hysteresis is large, the curvature of this waveform will be more apparent, while if the hysteresis is small, the waveform will be a better approximation of a triangle wave with more nearly linear slopes.

The voltage waveform on C1 is buffered to serve the triangle output by U1 which is an operational amplifier (linear) connected as a voltage follower. The buffering effect of U1 allows the triangle output voltage to drive a variety of loads without them affecting the stability of the oscillator.

There are a few integrated circuits that offer the combination of a comparator and an op amp in a single package, for example, LM392 and TLV2702.

With the same number of components but a different topology, a much more linear triangle wave can be generated. This requires connecting the operational amplifier with the capacitor in its feedback path, making it an integrator. This is detailed in Fig. 3.20.

In Fig. 3.20, the comparator U2 output square wave drives the input of the integrator, an operational amplifier at U1. To analyze the sequence of events, start with the output of the comparator in the "high" state. Since the integrator is an inverting stage, the high input will drive the output of the integrator in a ramp toward a low level. When that ramp reaches the lower point of the comparator's hysteresis range, the comparator will switch to a low output.

Fig. 3.20 RC oscillator with more linear triangle output

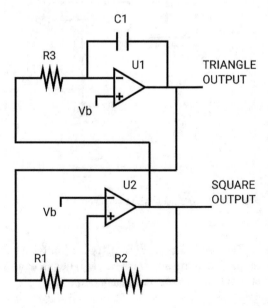

Note that the comparator has been changed from inverting in Fig. 3.19, to non-inverting in Fig. 3.20, to function properly with this oscillator. The low output of the comparator then drives the integrator to produce a rising ramp output, that will eventually reach the upper hysteresis point of the comparator. At that point the comparator will switch back to a high output state, and a new cycle begins.

To have a symmetrical waveform, the circuit of Fig. 3.20 requires that the bias voltage, Vb, be set to the mid-point of the comparator output logic levels. Minor adjustments of symmetry are easily made by changing Vb, but larger variations in symmetry are best accomplished as described next. The reason for this is to avoid the limits of the common-mode input voltage ranges of the op-amp and comparator.

Up to this point both the RC oscillators have supplied a symmetrical output, that is a 50–50 duty cycle and a symmetrical triangle wave. If it is desired to have an asymmetrical output, that is a pulse wave train with a duty cycle of other than 50%, or an asymmetrical triangle wave where the positive slope differs from the negative slope, often called a sawtooth wave, there is a simple addition to accomplish that. The half-cycles of the oscillators in Figs. 3.19 and 3.20 are equal length because the voltage across R3 is the same, while changing polarity. That is still true in Fig. 3.21, but a pair of diodes is used to split R3 into R3 and R4.

In Fig. 3.21, the voltage applied to R3 and R4 is still the same, but when its polarity switches, the diodes apply this voltage to R3 or R4 separately allowing the charging current to C1 to be different for the two half-cycles. This causes the

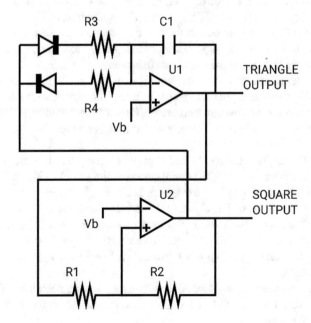

Fig. 3.21 Asymmetrical RC oscillator

duration of each half-cycle to be controlled by R3 and R4 separately. The changes applied to Fig. 3.20 that are shown in Fig. 3.21 may also be applied to Fig. 3.19 with a similar result.

There are more oscillator circuits shown in Chap. 5 Application Details, section "Multiple Comparator Circuits," "Voltage Controlled Oscillator," Fig. 5.6 and in Chap. 6 Logic Elements, the section titled, "Logic Elements as Oscillators," including an RC oscillator examples utilizing a Schmitt trigger inverter, Fig. 6.2 and non-hysteresis inverters, Fig. 6.3.

In the following section, "Timer IC", with Fig. 3.26, there is another oscillator, an astable multivibrator, and a voltage-controlled version is discussed.

3.1.7 Switch De-Bouncers

Interfaces between either humans or mechanical systems with digital electronics often depend on an electrical switch operation, closing and opening. In these situations, there is almost always a potential problem, particularly with the closing of the switch, and sometimes with the opening as well, where there is a "stutter" in the transition. This is called, "switch bounce".

In an electromechanical switch, such as a push-button, a toggle switch, a keyboard, a mechanical limit switch, even an electromechanical relay, or a similar device where two metallic contacts come together to establish a current path, at the moment of contact, there is almost always a mechanical bounce that causes the current flow to start and stop several times over a few milliseconds. This "stutter" effect could mislead a digital circuit connected to the switch to interpret that several events had occurred, instead of the single one intended. There are several circuits employing a comparator that can eliminate the switch bounce problem.

One of the most effective ways of dealing with switch bounce is to consider the bounce to be a short-term phenomenon and follow the switch with a lowpass filter that has a response time constant that is a bit longer than the bounce. Then the output of the low-pass filter needs to be converted to a clean logic-level transition by a comparator.

In Fig. 3.22, the components R1 and C protect the input of the comparator from electrostatic discharge that might be introduced at the switch. This circuit arrangement has a rapid discharge when the switch is closed, followed by a slow recharge when the switch is opened. The resistors R2 and R3 are shown connected to provide positive feedback for hysteresis, as discussed earlier. Hysteresis further mitigates against multiple bounces of the switch. This circuit is drawn from the viewpoint of a single-supply situation, making it necessary for resistor R2 to be returned to a mid-supply voltage.

Component values that could stand as a starting point for design are: C = 220 nF, R1 = 10 kΩ, 10 R1 = 100 kΩ (for R1*C = 2.2 ms), and R2 = 10 kΩ, R3 = 100 kΩ (for 10% hysteresis).

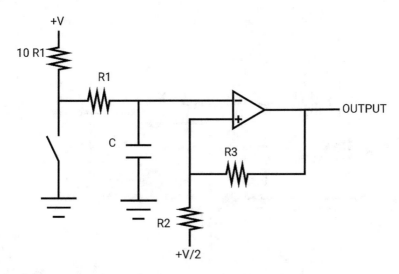

Fig. 3.22 De-bounce with grounded switch

Since most switches exhibit bounce behavior durations of less than 10 ms, the time constant of (10R1+R1)C should be at least 10 ms. Then the response time from operating the switch to the level change at the output of the comparator will be on the order of R1*C, likely one or two milliseconds. This gives near-instant response (on a human scale) while protecting against later bounces for 10 ms or more. This also means that repeated operation of the switch will be ignored unless delayed by well over 10 ms, likely a desirable feature, as it will avoid a shaky hand from creating two clicks unintentionally.

If it is more convenient to connect the switch to the supply voltage rather than ground, then the circuit in Fig. 3.23 operates in a fashion very similar to the circuit in Fig. 3.22. One difference in Fig. 3.23 is that when the switch is closed, there is a rapid charging of C through R1, followed by a slow discharge of (10R1+R1)C when the switch is opened. With component values in Fig. 3.23 set to the same as in Fig. 3.22, this arrangement performs the same response time patterns as the circuit in Fig. 3.22.

Another difference between Fig. 3.22 and 3.23 is that when the switch is activated, the output of Fig. 3.22 will transition from low to high, while the Fig. 3.23 output will transition from high to low state.

Switches that need to be de-bounced are generally SPST (single pole single throw), NO (normally open). However if a normally closed switch is actuated and goes open, then there is normally little or no bounce phenomenon. This is an attractive situation, but it incurs the disadvantage of greater power consumption because the closed switch must be conducting a current that is interrupted when the switch is actuated. Still, the circuits described above are applicable to normally closed switches as well.

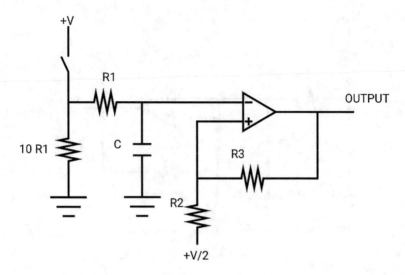

Fig. 3.23 De-bounce with the switch to supply

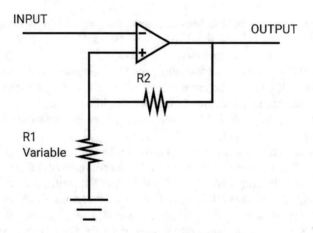

Fig. 3.24 Phase shifter

3.1.8 Voltage-Controlled Phase Shifter

The deceptively simple circuit in Fig. 3.24 can perform a very useful function that is otherwise difficult to achieve. When a comparator is fed with either a sine wave or a triangle wave (as from Fig. 3.19), and is implemented with hysteresis, the phase shift through the stage can be adjusted by changing the value of a resistor in the positive feedback path that varies the hysteresis. Refer to Fig. 3.1, and consider that if the inverting connection is used, then the input will be connected to the inverting input of the comparator, and the left end of R1 will be grounded, making it convenient for varying. This is detailed in Fig. 3.24.

In Fig. 3.24, the variable resistor can be any of a manually variable potentiometer, a field effect transistor whose channel resistance is modulated by the voltage applied to the gate, or a multiplying digital-to-analog converter (DAC), which then allows the phase to be adjusted by a digital control word. For direct digital control of the phase shift an integrated circuit digital potentiometer such as the Texas Instruments TPL0401 may replace R1 alone, or possibly R1 and a portion of R2. When a digital potentiometer is used, its capacitance may need to be compensated by a capacitor in parallel with R2.

If the input signal is a triangle wave, then the phase delay will be linearly related to the resistance of the variable resistor, with this linearity depending on the precision of the triangle wave. Even with the slopes of the triangle being portions of the exponential charging slope of R3 and C1 in Fig. 3.19, phase shift ranges of more than one radian may be accomplished with useful linearity.

When more precise linearity of phase shift versus resistance change is required, the triangle wave may be generated by an integrator or by current sources charging and discharging a capacitor.

For a constant amplitude input to this phase shifter, a change in frequency will result in a constant number of degrees (or radians) of phase shift. This is due to the hysteresis choosing a particular amplitude on the wave to trigger the change in output. Correspondingly, for a change in amplitude, this phase shifter will show a change in phase shift due solely to the amplitude change.

An earlier section, Offset Effect of Hysteresis, discusses the fact that an offset voltage is incurred by the application of hysteresis. This is especially important to consider in the present case where the hysteresis may be large and variable, thus making the effective offset large and variable. As was mentioned in that earlier section, the offset effect may be minimized by making the reference voltage for the comparator equal to the mid-point of the two logic levels at the output of the comparator. For this phase shifter application, that is highly recommended, and topologies shown in Figs. 3.3 and 3.16 earlier provide a starting point.

3.1.9   Timer IC

The 555 timer IC, available from many manufacturers in both bipolar and CMOS versions, is a simple way of converting an analog signal to a pulse width modulated (PWM) digital signal, or in general an inter-pulse interval (IPI) modulation. If more of these functions are needed, the 556 is a dual version, and 558 is a quad. The original implementation of these circuits was in bipolar technology, but updated, improved performance versions are available in low power CMOS.

A major advantage of the CMOS versions is the avoidance of a major supply current spike in the bipolar versions that arises from the overlap of the on-states of the two output transistors in their totem pole structure. This required a significant

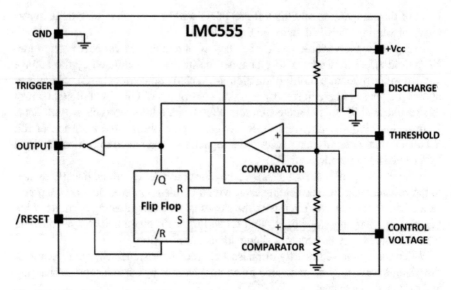

Fig. 3.25 Block Diagram of 555 Timer (Courtesy Texas Instruments)

bypass capacitor at the supply pin. The CMOS versions should still be used with a bypass capacitor at their supply input, but 100 nF is usually adequate, instead of the 1.0 µF needed by the bipolar circuits.

As shown in Fig. 3.25, the 555 contains two comparators, with external access to one input of each. The first of these comparators, which we will call the "reset comparator", has its output driving the R input of a reset-set (RS) flipflop, and the external input to its non-inverting input is labeled "Threshold". The second comparator, which we will call the "set comparator", output drives the S input of this same flipflop, and the external input to its inverting input is labeled "/Trigger". This flipflop also has a reset input, labeled /R, available externally. The Q output of this flipflop is available as a digital output having logic levels set by the supply voltage. The flipflop output also drives a transistor having a grounded emitter/source and an open collector/drain that is accessible externally, labeled "Discharge". For supply voltage, various versions of these devices can operate over the range of 0.9–18 V.

The internally connected inputs of the two comparators are fed with fractions of the supply voltage by an internal resistive voltage divider. The typical values of these two reference voltages are 1/3 of the supply for the set comparator, and 2/3 of the supply for the reset comparator, with tolerances of ±20% on each. Comparator input offset voltages are covered by these tolerances. The 2/3 supply node of the voltage divider is also made available on an external pin called "Control Voltage", which allows monitoring or modifying the voltage at that node, for example with a resistor to the supply or to ground, or being driven by a low impedance source. All three of these resistors have nominally the same value, which, in the bipolar version is 5 kΩ, but higher values are used in the CMOS versions.

If the Control Voltage port is driven to some voltage other than 2/3 of the supply, then, of course, the reference voltage for the set comparator is also affected, being nominally half the Control Voltage. This does allow the reference voltages of both comparators to be set independently of the supply voltage or at least decoupled from rapid variations of the supply voltage by connecting a bypass capacitor from Control Voltage to the ground.

The truth table for the 555 is unique in that no other device operates this way:

```
Reset/  Trig   Thres  Out
------  ------ ------ ------
   1      0      X     1 Trig Overrides Thres
   1      1      0     1
   1      1      1     0
   0      X      X     0 Reset/ Overrides Thres & Trig
```

This may be summarized by noting that Reset/ = 0 forces the output to 0; if Reset is high, then Trig 0 forces the output to 1; and if Trig is 1, then the output is Thresh inverted.

The 555 can function as a Schmitt trigger when both the Threshold and Trigger inputs are tied together and used as a single input. In this case, the hysteresis of the resulting Schmitt trigger is nominally one-third of the supply voltage, centered around half the supply voltage. The Schmitt trigger is discussed in more detail in Sect. 3.2 under Enhancements - Hysteresis. If it is desired to capacitor couple the input signal, then a resistor from this input should be connected to half the supply voltage, or alternatively, two equal resistors, one to the supply and one to ground.

If only the Threshold input is desired to be capacitor coupled, then a single resistor from the Threshold input to the Control Voltage port provides appropriate dc bias. See R3 in Fig. 3.3 for a similar connection. Either of these capacitor coupled arrangements implements the ac coupled average tracking discussed in more detail in Sect. 3.3 under Decision Level Setting - Adaptive Thresholds.

Using the 555 devices to convert from the analog domain to the time domain has a limitation on the bandwidth of the analog signal. Since the toggle frequency of these 555 devices is limited to the low megahertz range at best, then they are limited in the bandwidth of the analog signal they can process to the audio range, or the low hundreds of kilohertz at best.

Figure 3.26 is an example of using the 555 as a PWM. A synchronizing trigger is applied to the/Trigger input, the Threshold and Discharge inputs are tied together and to a timing capacitor to the ground as well as a timing resistor to the supply. This creates a one-shot or monostable multivibrator (MSMV). Then the analog voltage is applied to the Control input and it controls the pulse width.

The 555 may also be connected to operate as an astable multivibrator, and when an analog voltage is applied to the Control input, it operates as a pulse position modulator (PPM) or pulse frequency modulator (PFM).

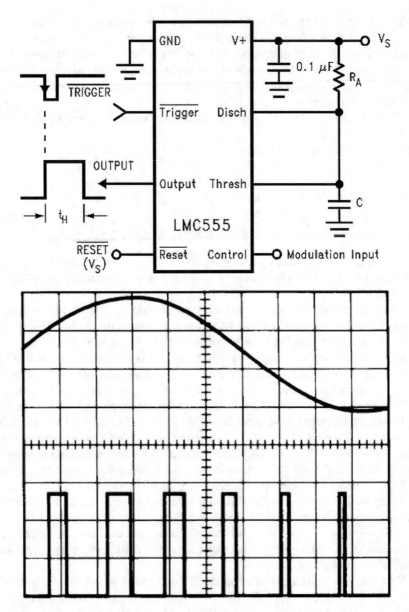

Fig. 3.26 Pulse Width Modulator with representative waveforms. The upper trace is the modulating input, and the lower trace is the resulting output pulse train. (Courtesy Texas Instruments)

3.1.10   Voltage-to-Frequency Converters

There are a number of circuits that perform the voltage-to-frequency conversion (VFC) function, both in comparator application notes and as complete self-contained circuits. This is the same function mentioned earlier as the pulse

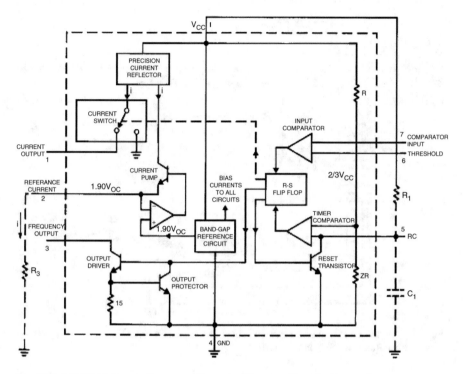

Fig. 3.27 LM331 Voltage-to-Frequency Converter (Courtesy Texas Instruments)

frequency modulator (PFM). Such circuits can simply provide an analog-to-digital conversion with linearity of the order of 100 parts per million. This is roughly the precision afforded by a traditional ADC of 13 to 14 bits. One such example is the LM331 in Fig. 3.27.

The LM331 contains two comparators, both of which are pivotally important to the functioning of this circuit. While the characteristics of the comparators are not separately specified, the overall specifications of the device cover their operation.

Later in the Sect. 5.4.2 Voltage Controlled Oscillator, Fig. 5.7, there is a similar appearing circuit, but the operation of these two circuits is quite different, starting with the fact that the output frequency of the LM331 is proportional to the input voltage, while in the circuit of Fig. 5.7, the output frequency decreases with an increasing voltage input. Further the LM331 is limited to a 100 kHz maximum frequency, while the circuit of Fig. 5.7 can be designed to operate to 10 MHz and beyond.

3.1.11 Multiplexed Comparators

There are comparator circuits with useful support circuitry surrounding the comparator proper, such as multiplexed inputs and gated outputs. Many microcontrollers include such a feature.

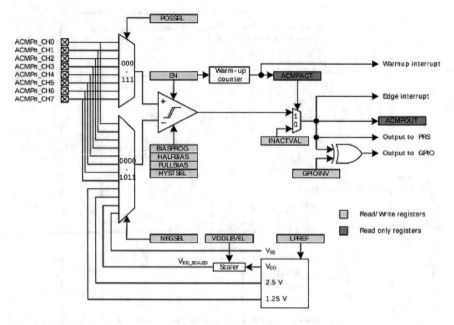

Fig. 3.28 Multiplexed Comparator Portion of the EFM32. From Silabs AN0020 (Courtesy Silicon Laboratories) Reference Silicon Labs (2013)

An example of this type of circuit is the analog comparator function in the microcontroller Silabs EFM32. The comparator portion of this product is shown in Fig. 3.28. There is a multiplexer for eight external inputs, each of which may be switched to either the inverting or non-inverting inputs of the comparator, along with four more inputs from internal voltage sources to the inverting input of the comparator. These four internal sources are Vss, Vdd scaled, 2.50 V, and 1.25 V. The scaling of the Vdd is controlled by a six-bit binary word. This is a very low power product, and it allows programming the bias current to the comparator over a 2500: 1 range; with corresponding inverse effects on response time of the comparator, but the response times are not specified.

References

Trump, Bruce (2013) Comparators—what's all the chatter? https://e2e.ti.com/blogs_/archives/b/ thesignal/posts/comparators-what-s-all-the-chatter. Accessed 2021 Oct 23
Silicon Labs (2013) Analog Comparator AN0020 https://www.silabs.com/documents/public/ application-notes/AN0020.pdf. Accessed 2021 Oct 23

Chapter 4
Nonideal Behavior

4.1 Dispersion, AM-to-PM Conversion

When the signal to be processed by the comparator is relatively narrowband, that is its sidebands are substantially contained within an octave of bandwidth which might be from (2/3)fc to (4/3)fc, where fc is the center frequency or carrier frequency, then the signal may be considered in the time domain to be generally sinusoidal, at least over one or a few cycles. For these generally sinusoidal signals where the information they carry is contained solely in their phase, not amplitude, like BPSK, quadra-phase shift keying (QPSK) or eight-phase shift keying (8PSK), the comparator can perform the very useful function of stripping off or ignoring amplitude variations. This action can improve the signal-to-noise ratio and can make automatic gain control (AGC) either unnecessary or much less critical. However, if amplitude variations presented to the comparator result in changes in the propagation delay through the stage, then an undesirable conversion has occurred that can be called AM-to-PM conversion. This parameter is shown on comparator data sheets as dispersion.

When the signal level presented to a comparator varies over more than a small range, ±1 dB for example, then the propagation delay of the comparator stage itself must be expected to vary. Clearly, that may be viewed as a changing phase shift in the path. This AM-to-PM effect is almost always a nonideal, undesired, source of error or degraded system performance.

While the dispersion characteristics of a comparator can obviously result in degradation of the AM-to-PM performance of a signal path, beyond this, there are many other cases where dispersion injects error or randomness. An example of this is when an audio bitstream is being processed. Consider a 5 kHz tone with an amplitude change of 10 dB that incurs a dispersion time shift of 2 μs. This would occur with a dispersion factor of 200 ns/dB, relatively poor performance, near the worst in Table 4.1. Since this is 1.0% of the period of the fundamental, it will likely create near 1.0% harmonic distortion.

© The Author(s), under exclusive license to Springer Nature Switzerland AG 2022   55
M. C. Fischer, *Comparators*, https://doi.org/10.1007/978-3-030-95742-1_4

Table 4.1 Dispersion of a range of comparators

Manufacturer	Part Number	AM-to-PM
Hittite	HMC674L	0.4 ps/dB
National Semi	LMH7324	0.5
Maxim Integ	MAX9600	1.5
National Semi	LMH7322, 24	2.2
Analog Devices	AD96685	2.5
Maxim Integ	MAX9691	7.5
Linear Tech	LTC6957	25
National Semi	LMH7220	25
	LM360	27
Texas Inst	TL3016	60
Linear Tech	LTC6752-2,3	64
Linear Tech	LT1394	\
Linear Tech	LT1720	\ All:
Maxim Integ	MAX90X	> 50
NXP Semi	NE521	/to 100
Texas Inst	TLV3501,2	140
	TL3116	150
National Semi	LMV7219	170
ST Microelectr	TS3011	500/350 -/+
National Semi	LM311	1000
National Semi	LMV7235,9	1250
Linear Tech	LT1671	1500
Intersil (RCA)	CA3089/CA3189	1600
National Semi	LMV761, 2	5000
Intersil (RCA)	CA2111	9000
	CA3012	no data
National Semi	LMV331,393,339	21000
Texas Inst	TLV7081	25000
Analog Devices	ADCMP370	70000
Linear Tech	LT1716	94000
Texas Inst	TLV7211	300000

In a system that contains a comparator, there should be an error budget that specifies for a given amount of signal-level change, a limit for how much delay change through the comparator can be tolerated. This may be quantified by starting with a number of decibels of signal level change to be expected, then deciding how much delay change can be tolerated, caused by the signal level change alone. Next divide the delay change in seconds, by the signal level change in decibels, to get the AM-to-PM dispersion factor limit for this one block in the system block diagram. This factor, with units of picoseconds per decibel, is the topic of the following discussion.

On the data sheets of comparator products this delay variation is called dispersion. Since the change in delay expressed in seconds does not seem to be sensitive to the frequency of the input, it is well to ignore the frequency and compare the specifications of comparators in terms of time shift in seconds rather than phase shift in radians or degrees versus amplitude.

When dispersion is plotted on a graph of linear time delay versus linear signal level, almost all comparators show a curved relationship that suggests plotting a linear time delay versus a logarithmic signal amplitude. Obviously the decibel makes a convenient and familiar unit for expressing changes in signal level on a logarithmic scale. For many comparators, this results in a much more nearly straight-line plot. By fitting the data with a straight line, this line may be seen to have a slope that can be stated in picoseconds/decibel. This one parameter may then be used to compare various comparator's performance with respect to dispersion. A representative assortment of comparators is shown below with their dispersion slopes. Note that the slower comparators may have huge slopes and even among the very fastest comparators there is a wide variation.

Dispersion is usually only given as a typical value and is defined as the change in propagation delay for a given change in input overdrive. Overdrive is defined as the excursion of the input signal beyond that level necessary to cause the comparator to change output state. Most comparators also exhibit a differing amount of change in propagation delay for positive transitions of the output versus negative transitions, a difference that can be in the neighborhood of double.

Since many comparators exhibit a near-linear plot of propagation delay versus overdrive when the overdrive is plotted on a logarithmic scale, this suggests a simple though crude way of expressing the dispersion as a single number; take the largest propagation delay and subtract the smallest propagation delay, then take the corresponding overdrives and express their ratio in decibels. Next divide the change in propagation delay by the decibels of overdrive change, yielding an estimate of AM-to-PM performance in units of picoseconds/decibel. This is crude because the actual curve of delay vs. overdrive can have slopes that depart from this broad average by 2:1 or more. However, given the large range of this parameter over the various parts available, it is a useful measure for comparisons between devices.

The concept of relating picoseconds of delay change caused by decibels of amplitude change is simple and useful when applied at the system level where an allowable budget for this non-ideality may be set. Then the limit may be applied when choosing the particular comparator device.

Since many comparators have varying slopes of dispersion, if this parameter is a significant part of the system error budget, then a candidate comparator must be studied in more detail than the simplified, typical slope number used in this discussion. In Table 4.1 the dispersion slopes appear under the column heading AM-to-PM, with units of ps/dB.

The data sheets for many comparators give no clue as to their dispersion. Table 4.1 lists a few examples, most based on typical performance, not guaranteed specifications. This shows the greater than seven orders of magnitude of the range of this parameter. Note that generally, the lower dispersions occur with the faster comparators, while the higher dispersions occur with the slower comparators. This is reasonable since a shorter delay time has less opportunity for change versus drive level.

The groupings in Table 4.1 are only to emphasize the similar parts versus the widely varying performance.

The parts listed as National Semi are now available from Texas Instruments, since TI purchased National Semiconductor. The parts listed as Intersil are available from the Intersil division of Renesas; Hittite Microwave, Linear Technology and Maxim Integrated have been acquired by Analog Devices.

The LMH7322 was calculated from the data sheet typical performance of dispersion versus slew rate because it showed a greater sensitivity to slew rate rather than overdrive. The LMH7220 data sheet shows curves of propagation delay versus overdrive, slew rate, common-mode voltage, and temperature. Since am-to-pm is the result of the combined effects of dispersion due to slew rate and overdrive, both were combined to rate the LMH7220.

The TL3116 has maximum specs over temperature, which gives more confidence than the typical-only specs of most other devices.

As mentioned earlier in the section, "Comparator versus operational amplifier," it can be convenient to use an operational amplifier as a comparator. When this is done, the dispersion encountered may become quite surprisingly large. An example of this effect is the operational amplifier family TLV9061 (single), TLV9062 (dual), and TLV9064 (quad) from Texas Instruments. This is an unusual part in that the data sheet offers typical performance of the device used as a comparator and has curves showing the changes in propagation delay versus overdrive voltage. The result of analyzing these curves with the technique described above is an AM-to-PM performance of 125,000 ps/dB! This is primarily an operational amplifier and its dispersion is so poor that it is not included in the table.

Since this dispersion is 125 ns/dB, and the TLV7211 is 300 ns/dB, parts of this class should not be considered whenever dispersion is an issue, but still might be useful for something like zero-crossing detection of a 60 Hz power line signal. Such an application is shown later in the section, "Logic Elements as Comparators," with Fig. 6.1 employing a logic element where an operational amplifier might also be considered. Instead of the logic element shown, using an operational amplifier for this application would require the addition of some means of supplying a reference voltage to the other input of the operational amplifier's differential pair of inputs. If the dispersion is not a problem, then using an operational amplifier can offer much more tightly controlled specifications for input offset voltage and current, input bias current, input noise, and the threshold voltage of the stage.

4.1.1 Comparator Design for Low Dispersion

At the circuit level, another aspect affecting dispersion is the structure of the differential stage. Referring to Fig. 1.3, if the resistor supplying current to the two emitters is replaced with a current source, then the sensitivity of delay versus drive level will be reduced. Then if the current source has a low sensitivity to the voltage changes presented to it, this will further improve dispersion.

4.1.2   Limiting Amplifiers

The Intersil parts in Table 4.1, though no longer manufactured, were in very wide usage as limiting IF amplifiers for high-quality FM receivers, even though their AM-to-PM performance was dismal. (The CA2111 was listed as a replacement for the ULN2111 and the MC1357. There was a family of related parts, all with similar structure internally and quite similar specifications, most now considered obsolete.)

4.1.3   Dispersion Angle versus Time

Certainly, in a given system, the dispersion of a comparator may be seen in terms of unwanted phase shift in angle at the operating frequency. Once the system error budget or shift limit is established in angle terms, this parameter can be easily converted to its equivalent time shift over a specified amplitude change as in the equations below:

$$\Delta t = \left(\Delta\varphi(\text{radians})\right)/\left(2\pi f\right) \tag{4.1}$$

$$= \left(\Delta\varphi(\text{degrees})\right)/\left(360 f\right) \tag{4.2}$$

$$\Delta A = 20\log\left(V1/V2\right) \tag{4.3}$$

$$\text{Dispersion slope}\left(s/\text{dB}\right) = \left(\Delta t\right)/\left(\Delta A\right) \tag{4.4}$$

where:
f = The frequency in Hz of the signal applied to the comparator
$\Delta\varphi$ = The change in phase angle of the desired signal at the operating frequency, in degrees or radians as shown above
V1, V2 = The maximum and minimum amplitudes of the desired signal in V
Δt = The change in delay in seconds allowable due to a change in amplitude from V1 to V2
ΔA = Twenty times the logarithm to base 10 of the ratio of the maximum signal amplitude, V1, to the minimum signal amplitude, V2, in decibels
These parameters can then be applied to the selection of an adequate comparator.

4.1.4 Dispersion Slopes, Overdrive Versus Slew Rate

The preceding discussion is based on fitting the dispersion data with a straight line having a single slope. This is a more or less crude approximation mainly useful for ranking comparators versus their competition.

Looking more closely at a particular comparator often shows that the value of the dispersion slope can exhibit differences of 2:1 or more over various ranges of input signal amplitude. It follows then that it is important to properly specify the input signal level. The data sheets for comparators that give information about dispersion almost always specify the input signal level as "overdrive" in volts. Overdrive is defined as the input signal swing from the point at which it crosses the threshold voltage to the maximum that the input signal reaches.

Next, if we compare this concept of overdrive with a typical case of a more or less sinusoidal signal applied to a comparator, then the value of overdrive corresponds to the peak value of the sine signal (half of the peak-to-peak). This assumes that the threshold level of the comparator is set at the mean of the input signal, an ideal case, but useful for defining the units to use. This definition of overdrive is coordinated with that used on data sheets; however, there is reason to believe that the effective overdrive in most applications will be a small fraction of the peak value. This will be discussed in more detail in the following paragraphs.

A further consideration of the differences between a typical application and the overdrive response data is the fact that the overdrive is applied with a high slew rate, or alternatively, a short rise time with respect to the propagation delay of the comparator. In contrast, in most applications, the slew rate of the input signal to a comparator is likely to be much smaller, even orders of magnitude lower, than the overdrive signal to which the data sheet refers.

This can result in the comparator responding as if the "overdrive" is a tiny fraction of the peak value of the actual input signal, the result being an unexpectedly long propagation delay, and unexpectedly large variation with signal amplitude.

Some of this extended propagation delay can be shortened by the application of hysteresis. However, it must be appreciated that the hysteresis signal cannot begin to help the input respond until the output has begun to move and this delay can be the majority of the propagation delay without hysteresis.

Because of this slew rate situation, it is wise to assume that the actual propagation delay of a comparator will be that with a very small overdrive, unless the input signal has the unusual characteristic of a high slew rate.

All this means that comparators in realistic applications are responding to the slew rate of their input rather than to overdrive, and accordingly, it would be most useful to designers if comparator data sheets gave propagation delay versus input slew rate, as a few are beginning to do.

It may be that propagation delay data is given for varying overdrive because that test is easier to run, ignoring whether the data correspond to any real application. This lack is beginning to be covered by vendors. National Semiconductor gives data for dispersion versus overdrive, slew rate, and common mode, as well as several other parameters, for the LMH7220, LMH7322, and LMH7324. The LMH7322,4 exhibits the curious behavior that, over about a 20 dB range of overdrive, increasing overdrive actually causes a small increase in propagation delay. This is especially curious in view of the fact that, over a similar range of slew rate increase, the propagation delay decreases, as is more common.

On the other hand, an earlier product, the LMV7219 offers only overdrive dispersion data, and it is an order of magnitude worse than later products where more attention was given to that performance. Specifically, the LMV7219 incurs 3 ns of dispersion over an 18 dB range of input, for a slope of 167 ps/dB.

In contrast, the venerable old LM360 offers no plots, but an early data sheet shows typical data points with an input of a 10 MHz sine wave at 30 mVpp and 2 Vpp, a 36.5 dB difference, resulting in a 1 ns change in propagation delay (what happens in between?) that computes to a respectable 27 ps/dB dispersion slope, where it is listed in the table above. The current data sheet has none of those points and instead mentions that the typical delay varies 3 ns for overdrive changing from 5 to 400 mV. This represents 38 dB and dividing the 3 ns by this gives 79 ps/dB.

Since there is about a 3:1 difference in these two characterizations, it is well to notice that the input overdrive ranges for the two cases are quite different. While it may not be justified, it is tempting to surmise that for overdrives ranging above 30 mV, there are about 1.0 ns of dispersion, while below 30 mV, down to 5 mV there might be an additional 2 ns of dispersion. Other comparators that have more complete curves often show this type of behavior, in that the dispersion slope worsens at low signal levels, and flattens out at higher levels.

Since it would be useful to have a way of relating the slew rate of a real signal to the overdrive characterization seen on most data sheets, we can hypothesize how they may be related. Clearly any such relationship can only be verified through extensive testing. But a relationship that offers at least a rough approximation may help set expectations for actual behavior. A candidate hypothesis for relating slew rate to overdrive is the following:

If we start with the propagation delay of a comparator with a large overdrive, then this time interval may be considered the period during which the slewing of the input is determining the eventual delay. For example, if a comparator has a propagation delay of 5 ns with a large overdrive, then with a smaller, slower slewing signal, it may be the amount of voltage change during the first 5 ns after the input crosses the comparator threshold that determines the delay. Following this hypothesis, an input signal that is a 10 MHz sine wave of 1.0 V_{pp}, or 0.5 V_{pk}, has a slew rate of $\pi f V_{pk} = 15.7E+6$ V/s in the vicinity of its zero crossing. During a 5 ns interval this signal slews 80 E−3 V or 80 mV, so the 5 ns comparator may respond to this as if it were 80 mV of overdrive, even though the peak excursion of the signal is 500 mV.

For the example case above where the slew rate is 15.6 E+6 V/s, it is interesting to note that the plot of slew rate dispersion for the LMH7322 extends only down to 100 E+6, and up to 1000 E+6 V/s.

While the highest speed comparators may not be needed for zero-crossing detection of slow-moving signals, that area of application is also important. Examples where there are many millions of products are smart electric energy meters. In that case, the waveform to be sensed is 100 V rms (to 240 V rms or more) at 50 Hz or 60 Hz. In some of these units, the error contribution of the comparator needs to be kept below one microsecond. While the excursions beyond the supply rails will be clipped by diodes, the slew rate of the zero crossing is $\pi\,141V_{pk}*50\,Hz = 22\,E{+}3$ V/s. Note this is three orders of magnitude lower than the 10 MHz case above. Here we have somewhat less than 10 dB of signal variability and 1 µs of dispersion budget, which yields a dispersion slope limit of 100 ns/dB. In such an application, it may be appropriate to use comparators with some of the highest dispersion characteristics, National Semi LMV331, 393, 339 at 21 ns/dB, because of their otherwise desirable features and minimal cost. An example of a power line zero crossing detector utilizing a Schmitt trigger logic element is shown later in the Sect. 6.1 Logic Elements as Comparators with Fig. 6.1.

4.2   Noise Measurement

The noise figure of a receiving chain will have likely been determined by stages preceding the comparator, due to their gain presenting a higher level signal to the comparator. Still, there are signal conditions and applications where comparator input noise may be a concern, and since this is an unspecified parameter, there may be a requirement to measure it.

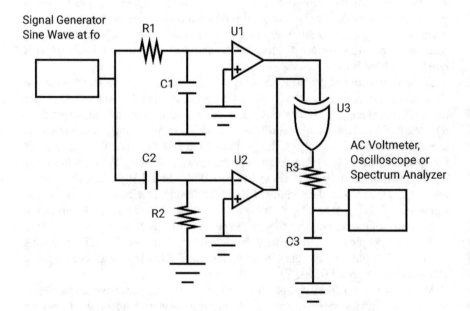

Fig. 4.1 A comparator noise measurement setup

If it is desired to measure the equivalent input noise of a comparator, there is a procedure to do so, with a setup as shown in Fig. 4.1. This measurement must be done indirectly due to the strongly nonlinear nature of the comparator's response. The results are discussed as phase noise at the output because the additive noise at the input results only in time perturbations of the zero crossings, which then appear as phase noise at the output.

This method of measurement of comparator input noise requires the measurement of two comparators at a time, U1 and U2. The noise output of the test system will be the combined noise of the two comparators. Assuming that the noises are uncorrelated, and equal, then their variances will add giving a result that is larger than that for each comparator by the square root of two. To solve for the noise of each comparator alone, a series of three measurements, pair-wise, among three comparators must be made. The procedure and computation for this are discussed later in this section.

To provide a low phase noise test signal, the sine wave signal generator should be a high-quality RF generator, or a quartz crystal oscillator with a sinusoidal output, for example. The low phase noise sine wave is split into +45° (R2, C2) and −45° (R1, C1) components and fed to the two comparators under test, U1 and U2, whose outputs will then be 90° apart. This is accomplished by making the resistor R1 equal to the reactance of the capacitor C1. The two resistors may as well be equal, as is true for the capacitors. Further, if the source is 50 Ω, then the resistors can be 50 Ω as well, since the reactance of the capacitors will add.

These two comparator outputs are then compared in a phase detector, a need that is well served by an exclusive-or gate, U3. For an active logic gate, in TTL the Schottky and AC CMOS families have been found to have low additive phase noise. Exclusive-or examples are 74S86, 74AS86, 74F86, 74LS86 74ALS86, and 74AC86. To work with comparators having ECL outputs, the MC100EP08 or NB7L86M are examples. A passive mixer can also be used as an extremely low noise phase detector if it is the double-balanced type. In most cases, such a mixer will have 50 Ω inputs and expect no direct current (DC) component, so the interface between the comparators under test and the mixer should be alternating current (AC) coupled and may need to include a series resistor.

The choice of test frequency for noise measurement deserves some consideration. Lower frequencies will have a slower slew rate that can give a more precise measurement of low noise levels. However, the test frequency places an upper limit on the frequencies of the noise spectrum that can be measured. This occurs because the output of the exclusive-or gate is a near-square wave of twice the frequency of the test signal. As a result, the spectrum of the noise is visible only up to somewhat less than the test frequency.

The spectrum of the comparator noise is of interest because it will consist of a region of flicker noise, that is $1/f$ or f^{-1} shape, at low frequencies, followed by a white spectrum (f^0, flat) at higher frequencies. The intersection of these two processes is called the "flicker corner." For bipolar transistor input stages of the comparator, the flicker corner can be from a few hundred hertz to a few thousand hertz. However, for MOS transistor input stages, the flicker corner will likely be from

10 kHz up to a megahertz. Clearly then, we would like to have the test frequency high enough to discern the flicker corner. For a broad application test setup, 1.0 MHz may be a good choice for the test frequency, while for testing very low noise bipolar comparators, we might choose 10 kHz or lower.

Representative component values are R1 and R2, 50 Ω, C1 and C2 are capacitors having a value that gives them a 50 Ω reactance at the measurement frequency, f_o. R3 is not critical, but 1000 Ω is reasonable, and C3 should have a reactance of twice the value of R3 at the highest frequency of noise that is to be measured.

The results are shown as phase noise or jitter at the output because the only effect of additive noise at the input is to cause time perturbations of the zero crossings, which then appear as phase noise at the output.

Calibration for a test system of this type may be accomplished by feeding two sine waves near the frequency of interest, but slightly different frequencies, into the two comparators to be measured, with their outputs driving the exclusive-or gate functioning as a phase detector. The low-pass filter after the exclusive-or gate will pass the difference frequency, and its waveform should be a triangle wave. The peak-to-peak amplitude of this triangle wave represents π radians of phase shift.

This occurs because the two square waves from the two comparators are sliding past each other in time, and the exclusive-or of these two signals, when low-pass filtered, will exhibit a linear ramp up for one-half cycle of the beat frequency, then a linear ramp down for the next half-cycle. The peak-to-peak amplitude of the triangle wave will also be the difference of the two logic levels, high and low, from the exclusive-or gate, that may be measured with an oscilloscope. For initial design considerations, the data sheet typical values for V_{oh} and V_{ol} will be indicative, but not exact due to the loading of the noise measurement circuit being different from the data sheet conditions.

For a phase detector noise output of V_n volts rms (root-mean square), and a calibration measurement of V_{cal} volts peak-peak, the noise measurement φ radians rms is calculated:

$$\phi = V_n \frac{\pi}{V_{cal}} \tag{4.5}$$

This is, of course, a broadband, total noise measurement, where the measurement bandwidth is determined by the low-pass corner of R3 C3 in Fig. 4.1.

Since it may be of greater interest to express this total noise in terms of additive jitter, the noise measurement of φ radians rms is easily converted to a time jitter number τ seconds rms:

$$\tau = \frac{\phi}{2\pi f_0} \tag{4.6}$$

where f_0 is the frequency of the test signal being processed by the comparators. This value of τ seconds rms will be valid only for the frequency and amplitude where it is measured and computed, while the φ radians measurement may be applied to any frequency at the same amplitude of input signal to the comparators.

4.2.1 Why Radians Rather Than Seconds?

Characterizing the jitter in radians makes it independent of the frequency. Why this is true may be reasoned from considering that the input noise of a comparator is added to the sinusoidal test signal applied to the input. This noise voltage will push the slewing test signal up and down as it nears the threshold of the comparator, thus causing the transition to occur earlier or later.

The amount of this time shift clearly is proportional to the inverse of the slew rate, so it is not constant for different frequencies, even at the same amplitude. What is constant, for the same amplitude, is the number of radians of phase shift caused by the noise for various frequencies of test signal. A proof of this in the derivation that follows:

As was defined earlier under Enhancements >Hysteresis, the slew rate in volts per second of the zero crossing of the test signal of amplitude V_{pp} is:

$$\frac{dV}{dt} = V_{pp} \pi f_0 \tag{3.1}$$

Next, if we apply the comparator's input noise V_i (V rms) to this slew rate, we may express the result in terms of the time shift, τ:

$$\tau = \frac{V_i}{\dfrac{dV}{dt}} \tag{4.7}$$

Then substituting for dV/dt:

$$\tau = \frac{V_i}{V_{pp} \pi f_0} \tag{4.8}$$

Finally substituting the earlier equation with φ radians rms for τ:

$$\frac{\phi}{2\pi f_0} = \frac{V_i}{V_{pp} \pi f_0} \tag{4.9}$$

Reducing the algebra yields:

$$\frac{\phi}{2} = \frac{V_i}{V_{pp}} \tag{4.10}$$

Or:

$$\phi = \frac{V_i}{V_{pk}} \tag{4.11}$$

where V_{pk} is the _peak amplitude_ of the test signal, and φ is in radians rms.

This shows that the jitter in radians is not a function of the frequency of the test signal, but is driven by the noise level divided by the amplitude of the test signal. This is a very simple relationship that is straightforward to consider in the analysis of a system that contains a comparator.

4.2.2 Comparator Noise Measurement Example

After arranging the test system, the first step is calibration. Let's assume that the triangle waveform out of the exclusive-or gate has an amplitude of 4.5 V peak-to-peak = V_{cal}.

Further assume that the comparators under test have a fairly low level of input noise of 10 nV/rtHz. Assume this is a flat (white) noise level and that we are concerned with a 1.0 MHz noise bandwidth. That yields an input noise level of 10 μV rms, over the bandwidth. This is true for each comparator in the pair being measured, and we may assume that the noise processes in the two comparators are uncorrelated, so their variances add. This will give a measurement result that corresponds to 14 μV = V_i, combined.

For this example, we will drive the comparators with 100 mV peak = V_{pk} (200 mV peak-to-peak, or 71 mV rms).

Using the earlier equation relating input noise to output phase jitter:

$$\phi = \frac{V_i}{V_{pk}} \tag{4.11}$$

With V_{pk} = 100 mV pk, and V_i = 14 μV rms, then
φ = 14E−6/100E−3 = 1.4E−4 = 0.14 milliradian rms.
Applying the calibration formula:

$$\phi = V_n \frac{\pi}{V_{cal}} \tag{4.5}$$

Where we have φ = 1.4E−4 and V_{cal} = 4.5, solving for V_n (the expected measurement result) gives:

$$1.4E-4 = V_n \frac{3.14}{4.5} V_n = 2.0E-4 = 200 \, \mu V \, rms$$

That illustrates the signal levels that are typical with reasonably low noise comparators and raises the point that a low-noise amplifier of at least 20 dB gain may be required following the exclusive-or gate and low-pass filter in the measurement setup.

4.2.3   Jitter Added by Noise

In the application of a comparator to a system, the system error budget may be expressed in radians, but much more often the concern will be seen as the comparator adding jitter to the signal, and that jitter expressed in seconds rms; the symbol used here is τ, seconds rms. As shown in the equation above:

$$\tau = \frac{V_i}{V_{pp} \pi f_0} \tag{4.8}$$

This equation assumes the input signal to the comparator has a slew rate that is similar to that of a sinusoid of the same frequency, and has a frequency (mid-band) of f_0 Hz. Clearly, if the input signal has a widely varying level, then the jitter contributed by the comparator input noise will vary correspondingly, along with the am-to-pm effect of the dispersion characteristic discussed earlier. These two effects are separate, independent, and may be evaluated separately, then summed. Note that the jitter is a random variable, having an rms value, while the dispersion should be considered non-random, causing a shift in the mean. Should the frequency be varying widely, then the comparator's contribution to jitter as rms seconds will be least at the highest frequency, and worst at the lowest frequency, as may be seen from the above equation.

If the comparator under consideration has been tested and its noise contribution is known in terms of its jitter in rms radians, as defined in the Eq. 4.5 above as φ, then that value of φ may be scaled to a different signal level, V_{sig} in volts peak-to-peak, by the equation:

$$\phi_2 = \phi \frac{2V_{test}}{V_{sig}} \tag{4.12}$$

where V_{test} is the peak value of the test signal used to test the comparator, hence the factor of two is needed to put it in the same scale with V_{sig} in volts peak-to-peak.

For a time domain calculation, where the comparator noise is known as φ radians rms, and the signal level is V_{sig}, volts peak-to-peak, then the jitter in rms seconds is:

$$\tau = \left(\frac{2V_{test}}{V_{sig}}\right)\left(\frac{\phi}{2\pi f_0}\right) \tag{4.13}$$

where f_0 is the frequency in Hz of the signal, and V_{test} is the signal level at which φ was measured. The factors 2 cancel, leaving:

$$\tau = \frac{\phi V_{test}}{V_{sig}\pi f_0} \tag{4.14}$$

Again, V_{sig} is the peak-to-peak value of the input signal. This equation shows how dealing with jitter in the time domain is a bit more complicated. Of course if φ is measured with $V_{test} = 1$ V peak, then the calculation is one factor simpler.

It may seem unnecessarily confusing that V_{test} is expressed as the _peak_ value, while V_{sig} is in _peak-to-peak_ volts. There are two reasons for this. First, the signal into the comparator in its application system is likely somewhat random, and most readily characterized as its peak-to-peak value. Second, the test signal for the comparator is a clean, stable sine wave, making its peak value readily discernible, and doing so drops a factor of two out of the equations.

4.2.4 Hysteresis Can Increase Jitter

Should an application of a comparator require large amounts of hysteresis, it should be noted that, for an input signal with fairly narrow bandwidth, the waveform will be more or less sinusoidal, and the large hysteresis can push the actual thresholds well away from the zero-crossings of the signal. In this situation, the slew rate of the signal can be significantly less than $V_{pp}\pi f_0$ when the threshold is reached. When this occurs, the input noise of the comparator can cause much more time jitter than was predicted by the foregoing equations.

4.2.5 Noise Figure-of-Merit

Since the noise measurement φ radians rms is not affected by the measurement frequency, but can be expected to be proportional to the reciprocal of the input signal level, this suggests a broadly applicable noise specification for a comparator. The noise might be either measured at 1.0 V rms, or normalized to that level, and stated as φ radian*volts, which would then allow the designer to divide this specified number by the signal level to be applied, thus yielding the expected phase jitter contributed by the comparator noise, in radians rms. This also makes quite clear the noise penalty of small signals, and low slew rate signals, as well.

In setting up the measurement, the most challenging condition for the comparators under test will be making the input signal as small as the lowest level signal expected in the final application. This will make the measured noise larger to keep it above the instrumentation noise of the measuring equipment.

It is worth mentioning that the power supplied to the exclusive-or gate must be extremely low noise since supply noise shows up directly at the output, masquerading as comparator noise. For the quietest comparators, this might require the use of a battery supply.

If it is desired to use an analog double-balanced mixer as the phase detector, then the digital levels output by the comparators under test must be translated to be able to drive the typically 50 Ω inputs of the mixer. This may be accomplished, for TTL levels, by the 74S140, which is characterized for driving 50 Ω lines. Of course the translators will add their noise to the total being measured.

The main frequency component in the output of the phase detector will be twice the input frequency, but the component of interest is the low-frequency noise which indicates the phase variations. The phase noise of the input signal will be identical at both inputs of the phase detector, so it does not appear in the result.

The frequency at which this measurement is made is not critical, though it might as well be near the frequency of the intended application for the comparator, to be a realistic test. This is because the phase shift in radians caused by the comparator noise will be constant regardless of frequency. That is true because lower frequencies have lower slew rates, and a given noise voltage will perturb the time of the transition proportionately more, yielding the same number of radians as at a higher frequency.

Since each of the inputs to the phase comparator is driven by one of the comparators under test, then the noise of both comparators, which can be assumed to be uncorrelated, will be summed in the output. If we assume the noise contributions of each comparator to be equal, then the noise of one comparator is 3 dB less than the measured result. Each of the noise sources in the path adds noise to the result, but since these sources are uncorrelated, the variance of each source adds to the others, not the rms. That relationship shows up in the following paragraphs.

If the noise of a single comparator is desired, then a trio of measurements may be made pair-wise of three comparators. The individual noise of each comparator is φ_a, φ_b, and φ_c in radians rms; but since their variances add, then the three rms measurement results will be:

$$\phi_{ab} = \sqrt{\left(\phi_a^2 + \phi_b^2\right)}$$

$$\phi_{ac} = \sqrt{\left(\phi_a^2 + \phi_c^2\right)}$$

$$\phi_{bc} = \sqrt{\left(\phi_b^2 + \phi_c^2\right)}$$

Solving these three measurements for the rms noise of a single comparator yields the equation to be used to combine the measurements:

$$\phi_a = \sqrt{\left(\frac{\phi_{ab}^2 + \phi_{ac}^2 - \phi_{bc}^2}{2}\right)} \qquad (4.15)$$

Inspection of the formula shows that the two measurements containing the noise of unit "a" are added, then the noise contributions of units "b" and "c" are subtracted, leaving twice the noise of unit "a," which is why the division by 2 is needed.

4.2.6   Noise Spectra

If a low-frequency spectrum analyzer is used to measure the comparator noise in the arrangement of Fig. 4.1, it will likely be found that there are two spectral slopes. At the higher frequencies, the noise is likely to be white, arising from Johnson noise of bulk resistance in semiconductors and shot noise from the conduction across semiconductor junctions. At lower frequencies, it is likely that a rising noise level will be found, that being flicker noise, having a 1/f slope, found at the inputs of all semiconductor devices. High levels of flicker noise have at times been associated with greater defect density in semiconductor crystal structures, and even predictive of impending failure or higher failure rate.

If the measuring instrument is a spectrum analyzer, then the results may be expressed in radians rms per hertz of bandwidth, and the regions of flicker versus white spectra will be shown. If the measuring instrument is an AC voltmeter, then the total of all the noise will be shown. The bandwidth of this measurement is set at the low-frequency end by the low-frequency response of the AC voltmeter. At the high-frequency end, the upper bandwidth limit may be set by the high-frequency response of the AC voltmeter or by the low-pass filtering action of R3, C3 in Fig. 4.1. These bandwidth issues need to be coordinated with the bandwidths following the comparator in the target system to make the jitter measurement a useful predictor of system performance.

The noise spectrum will almost certainly include flicker or 1/f noise at the lower frequencies. Since the entire noise spectrum will be imposed on the operation of the comparator, the contribution of the flicker component deserves consideration. With the flicker noise having a spectral shape of 1/f, falling at higher frequencies, then at some frequency, the flicker and white noises are equal. This frequency is known as the flicker corner, and above it there is no need to consider the flicker component. Since the flicker noise dominates the spectrum below the flicker corner frequency, that is the region of interest.

Some very low power comparators have quite slow response times, and correspondingly slow rise and fall times of their outputs, thus indicating a low noise bandwidth. In these cases, the flicker portion of the noise spectrum may contribute enough to the noise total to be of concern. In contrast to this, most medium and high-speed comparators have noise bandwidths that are so wide that the energy contribution of the flicker portion is minor at worst. An example of how that can occur follows:

Assume a comparator has output waveform transitions of 5 ns rise and fall times. As will be discussed later, this implies a noise bandwidth of 70 MHz. If the flicker corner is quite poor at 1.0 MHz, then the noise power in the region from 1 Hz to 1 MHz (relative to the white noise floor) is the integral of $1/f$, which is $\ln(f)$ and the definite integral is 13.8. For the rest of the noise bandwidth, from 1.0 MHz to 70 MHz, the integral is 69E6. These quantities are noise power, so including the flicker spectrum raises the total by a negligible amount; and this is a fairly pessimistic case. With that reasoning, we will ignore the contribution of flicker noise in the discussions that follow.

The best bipolar transistor front ends of op amps (and presumably comparators) can achieve a flicker corner of 100 Hz or even lower. Most bipolar front ends have flicker corners between 1 kHz and 100 kHz. MOSFET front ends seem to exhibit more flicker than bipolar, having flicker corners from 10 kHz to 1 MHz.

There is an alternative noise measurement method, but it places the comparator in an unrealistic operating mode. This entails connecting a resistor from the output to the inverting input, and another resistor from the inverting input to ground. This causes the comparator to operate in the mode of a high gain operational amplifier. The ratio of the resistors sets the gain. The output noise may be measured and reduced by this gain factor to infer the equivalent input noise.

Although the equivalent input noise of a comparator could contribute significantly to the total noise of a system, almost all comparator products have no noise specification on their data sheets.

The noise of a comparator becomes combined with the signal and noise that is input to the comparator, and since the comparator noise may be considered uncorrelated with the noise accompanying the signal, their powers add. In cases where the signal has a very good signal-to-noise ratio, the noise of the comparator may cause a significant degradation. The noise of the comparator is added to the noise accompanying the signal ahead of the nonlinear decision process in the comparator. The effects of the comparator's action on the signal-to-noise ratio are dealt with in a later section, Hard Limiter Effects on Signal to Noise Ratio.

4.2.7 Noise Examples

To explore a practical example of comparator noise affecting the signal-to-noise ratio of the incoming signal, consider a small incoming sinusoidal signal of 100 mV peak-to-peak with a signal-to-noise ratio of +10 dB. This signal has an rms value of

35 mV rms, and the noise, being 10 dB lower, has an rms value of 11 mV rms. If the comparator has an input noise level of 100 nV/rtHz, and a bandwidth of 100 MHz, then the comparator's input noise contribution is 1.0 mV rms. To combine this with the noise accompanying the signal, since the two noises can be expected to be uncorrelated, and their noise powers add, then we square each of the noise voltages, add, and take the square root of the result. That is sqrt($11^2 + 1^2$) in mV rms for a result of 11.045 mV rms total noise, or a tiny fraction of a decibel degradation due to a fairly noisy though small signal, and a fairly noisy comparator.

This analysis raises the question of what noise bandwidth to use for the noise contributed by the comparator. Surely in many systems, the signal bandwidth will have been well controlled and defined before it reaches the comparator. But then the comparator has its own noise bandwidth, which is generally much wider, even by orders of magnitude. The noise bandwidth of the comparator may be estimated by considering the rise and fall times of its output waveform. Clearly this is not defini-tive, but it is indicative, and it may be the best indication we have versus a difficult direct measurement. To infer the noise bandwidth from the rise and fall times, the simplest approach is to use the well-known rule of thumb for simple circuits that the bandwidth, BW, is equal to 0.35 divided by the rise time, t_r:

$$BW = \frac{0.35}{t_r} \tag{4.16}$$

This formula is appropriate in Hz and seconds, or just as well in GHz and ns, while the rise time is defined as 10–90%. For example, the 100 MHz bandwidth men-tioned above implies a rise time of 3.5 ns. While this relationship is based on a Gaussian time response of a low pass filter, similar to the vertical amplifier of an oscilloscope, it is a useful approximation for any simple low-pass structure.

An example case where the noise contributed by a comparator could be critical is where a very low jitter clock signal sine wave is turned into a logic level square wave by a comparator. To compute the effects of a typical situation, start with a clock signal having a white phase noise floor of −160 dBc, a frequency of 10 MHz and an amplitude of 1 V rms. If this is applied to a comparator that has a very mod-erate 10 nV/rtHz of input noise, and a noise bandwidth that might well be 100 MHz, then this results in a broadband noise level of 100 μV rms. Applying this as the V_i term in the equation developed earlier:

$$\phi = \frac{V_i}{V_{pk}} \tag{4.11}$$

where V_i is the noise added by the comparator, V_{pk} is the _peak_ amplitude of the 10 MHz signal, and φ is in radians rms.

This results in:

$$\varphi = 1E-4 / 1.4 = 71E-6 \, radians \, rms$$

To express this as time jitter, consider that 2pi radians at 10 MHz is 100 ns. Then to convert radians to seconds, we multiply by 1E−7 and divide by 2pi, resulting in $\tau = 11E-12$ or 11 ps rms jitter.

To compare this result with the incoming signal white phase noise floor of −160 dBc/Hz, first we convert the decibel ratio to a power ratio which is $1E-16 = L_r$ in the next equation:

$$\sigma_y = \frac{\sqrt{L_r f_h}}{2.565 f_0 \tau} \tag{4.17}$$

This equation is from: Fischer (1976).

For this example, f_o is the 10 MHz signal frequency 1E7, and f_h is the bandwidth of the noise or signal path, in this case likely about 10 MHz or 1E7 and the τ is the averaging time of the measurement, which we will remove in the next step. Plugging those parameters into Eq. 4.18:

$$\sigma_y = \left(\mathrm{sqrt}\left(1E-16*1E7\right)\right)/\left(2.565*1E7*\tau\right)$$

Then to get σ_x, the jitter of the zero crossings, we multiply the above result by τ, thus removing it from the equation:

$$\sigma_x = \left(3.16E-5\right)/\left(2.565E7\right) = 1.2\,\mathrm{ps}$$

But recall that the comparator noise-induced jitter was 11 ps. This result shows that even a fairly low noise comparator can degrade the jitter of a fairly low noise, high-level source by almost 20 dB; a radically different result from the earlier example. Since there are now precision sources available that have a white phase noise floor of $L = -180$ dBc/Hz, which is 20 dB better than the example above, it is clear that the choice of a comparator can be quite critical. An example of this concern is addressed in the Reference Analog Devices (2017).

References

Analog Devices data sheet LTC6957 (2017) https://www.analog.com/media/en/technical-documentation/data-sheets/6957fb.pdf Accessed 2020 May 23

Fischer, Michael C. (1976) Frequency Stability Measurement Procedures, 8th PTTI, 1976, p 575-617; https://babel.hathitrust.org/cgi/pt?id=osu.32435058668229&view=1up&seq=657 Accessed 2021 Oct 15

Chapter 5
Application Details

5.1 Op-Amp Enhanced Comparator

Where absolute minimum noise at the input is a requirement, it may be worth adding a stage ahead of the comparator proper. This added input stage is an op-amp with its input equivalent noise having a specified limit. Further, it is desirable for this added stage to have enough gain to make the unspecified noise of the following comparator inconsequential. This raises the likelihood that the op-amp will saturate on large signal swings and degrade the response speed of the combination. To keep the op-amp in its high-speed linear region, some type of nonlinear feedback must be arranged. In other words, the op-amp stage must be configured to be what is sometimes called a soft limiter.

A soft limiter may be accomplished with a negative feedback path around the op-amp that contains a back-to-back pair of Zener diodes. While this illustrates the concept, Zeners have a number of disadvantages in this application. The best-performing Zeners have voltages of 5–10 V. This means that the limited output swing of the stage would be at least 10 V peak to peak. In many cases, this would be inconveniently large. Another issue is the rather large junction capacitance of Zeners, which would reduce the bandwidth of the stage.

Of course, a pair of anti-parallel light-emitting diodes (LED) may be used in the feedback path to yield a larger swing of the limited output, on the order of 4 V peak to peak. This can have the advantage of the LEDs glow indicating the presence of sufficient input signal, while a possible disadvantage could be the reduced bandwidth resulting from the higher capacitance of the LEDs.

Instead of Zeners or LEDs, the forward voltage of silicon PN diodes may be used. A pair of antiparallel diodes will result in an output swing of about 1.3 Vpp. Again the diode capacitance, while much less than Zeners, will reduce the stage bandwidth. A way of reducing this effect is to use two pairs of antiparallel diodes in series, with a resistor to ground between the two pairs, as is shown in Fig. 5.1. This will result in an output swing of about 2.6 Vpp. This signal to the comparator will

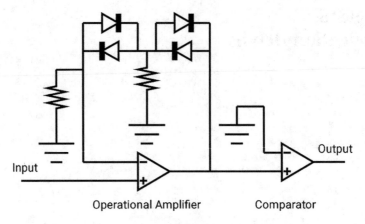

Fig. 5.1 A soft-limiting amplifier driving a comparator

also have a very greatly enhanced slew rate, usually several orders of magnitude more than the input.

In Fig. 5.1 an op-amp—comparator cascade is shown, with the op-amp having soft-limiting feedback to avoid saturation. An appropriate choice for the feedback may be the BAV99 antiparallel pair of diodes. Each diode has a maximum junction capacitance of 1.5 pF at zero bias. A pair of these in parallel will have 3 pF, and with 1.0 kΩ to ground, has a corner of 50 MHz. The grounds shown imply signal ground, likely a mid-supply point.

This circuit is shown with direct coupling for simplicity, while it will likely be convenient to ac-couple the input, and possibly the drive to the comparator. Should ac coupling be included, note that this implies an average-tracking adaptive decision level, as was discussed in the Sect. 3.3.1 Decision-Level Setting—Adaptive Thresholds. In that section, there are many alternative adaptive threshold topologies that are all applicable to this block. But note that the adaptive threshold voltage must be applied to the ground symbols in Fig. 5.1 in order to achieve the desired effect, because the threshold decision is being determined by both the operational amplifier stage and the comparator stage. This will likely require a voltage-follower stage to present a high impedance to the output of the adaptive threshold circuit, while offering a low drive impedance to all of the ground symbols in Fig. 5.1.

Since essentially all comparators exhibit variations in delay versus overdrive and slew rate of their input, as well as having no specification of input noise, then this cascade of a soft limiter driving a comparator can offer a substantial improvement in the overall performance. It can offer specified noise performance of the op amp and defined overdrive and slew rate, thereby bringing several error sources under design control.

In any system where there is a concern for the delay dispersion and/or the noise of the comparator, it is well to consider the soft limiter to drive the comparator. This concern would typically arise when the input signal to the comparator stage could have a dynamic range of 10 dB or more. As was seen in Table 4.1, a midrange comparator could have a dispersion of 100 ps/dB, which, for a 10 dB signal level change, becomes 1000 ps or 1.0 ns. Similarly, when the signal level drops 10 dB or more, the noise of the comparator becomes that much more of a factor.

The soft limiter itself can contribute some delay dispersion due to the recovery time of the diodes. This suggests the use of Schottky diodes, but these points have not been verified by experiment.

Since the circuit of Fig. 5.1 likely offers the best noise performance of any covered here, this is an appropriate point to raise the issue of the noise contribution of the resistors in it. The resistor between the two sets of diodes will not contribute to the equivalent input noise of the stage, because it is isolated by the diodes being in a nonconducting state when the input signal is crossing the threshold voltage. The diodes go into conduction only when the output swings high or low and their conduction provides feedback that keeps the operational amplifier from saturating.

While the input signal is crossing the threshold region, the noise of the resistor on the left side of the diodes will be added to the input signal by the operational amplifier. If that resistor has a value of 1.0 kΩ as suggested above, then its open-circuit Johnson noise will be 4.0 nV/rtHz at an ambient temperature of 290 K. Low noise operational amplifiers will have a similar amount of equivalent input noise, but the lowest noise units can get below 1.0 nV/rtHz, the noise of a 50 Ω resistor. These values are computed from the familiar Johnson noise equation:

$$V_n = \sqrt{4kTR} \tag{5.1}$$

V_n = RMS noise voltage per Hz of bandwidthwhere:
   k = Boltzmann's constant of 1.38 E−23 J/K
   T = Absolute temperature in kelvin
   R = Resistance in ohms

To combine the effects of the resistor noise with the operational amplifier noise, consider that the two noises are uncorrelated, and their powers add. That means that the two quantities of V/rtHz must be squared, added, then take the square root of the sum to get the combined noise. In the case of the 1.0 kΩ resistor and an operational amplifier with 4 nV/rtHz of equivalent input noise, the combined result is 5.7 nV/rtHz.

There are integrated circuits where a single package contains both an op-amp and a comparator, such as the TLV2702, and two such pairs, the TLV2704, from Texas Instruments. These parts are rather specialized as extremely low power, which results in their having quite low bandwidth of 5.5 kHz and noise of 500 nV/rtHz. Also, there is an IC that contains an op-amp, a comparator, and a voltage reference, the TC1026 from Microchip.

Because all the well-performing comparators have differential inputs, if the source of the signal to be processed by a comparator is balanced or differential, then it may be desirable to make the amplifier preceding the comparator one of the full differential type as shown in Fig. 5.2.

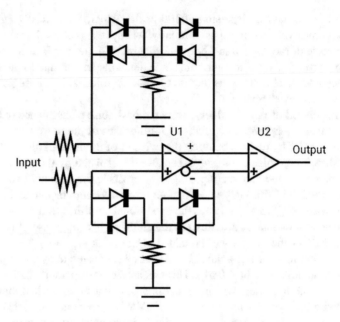

Fig. 5.2 Full-differential preamp for comparator

In Fig. 5.2, since the amplifier, U1, has both differential inputs and differential outputs, to preserve the amplifier characteristics, the negative feedback path with the four diodes and two resistors shown in Fig. 5.1 needs to be duplicated. In Fig. 5.2. one feedback path goes from the (+) output to the (−) input, while the parallel feedback path goes from the (−) output to the (+) input. This makes the amplifier stage symmetrical with a total of eight diodes and four resistors. This topology results in a total drive voltage to the comparator, U2, of $4\,V_{be}$ peak-to-peak, or about 2.6 Vpp, for silicon PN diodes. Note that over much of the input waveform, while some diodes are conducting, the input impedance of this stage is little more than the two input resistors. As a further note, it is not a straightforward task to add hysteresis or an adaptive threshold to this circuit. Of course, a comparator having built-in hysteresis can be utilized in this arrangement.

5.2   Detailed Circuit Example

All of the preceding diagrams have shown the comparators without power connections, without power bypass capacitors, and in a rather generic state to emphasize the signal paths while keeping the artwork as simple as possible. Clearly, a complete, practical comparator circuit must have many more components, and the following figure shows an extreme example of that:

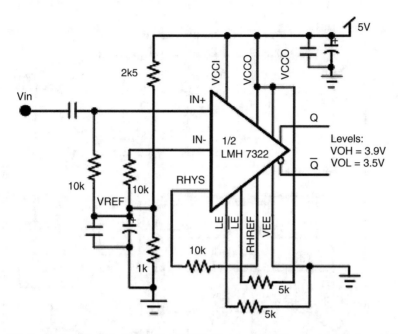

Fig. 5.3 Detailed application circuit of LMH7322 (Courtesy Texas Instruments) Reference Texas Instruments (2013)

In Fig. 5.3, the capacitor in series with the input line and the 10 kΩ resistor shunted to VREF comprise an input high-pass filter, as was discussed in the section Enhancements >Decision Level Setting—Adaptive Thresholds >Average Tracking, Fig. 3.3. This is a very high-speed comparator capable of toggling at 4 Gb/s, a propagation delay of 700 ps, and the dual capacitor bypass on the main supply and on the Vref are reflective of that. Each of these capacitors should have its own vias to the ground plane, not shared with any other component (unlike the way the schematic is drawn). The port labeled RHYS is a special port that allows external control of internally implemented hysteresis. The 10 kΩ resistor shown will result in about 10 mV of hysteresis.

5.3 Detailed System Example

A number of applications of comparators have been mentioned, and in this section, we will show many of the details of a portion of a communication system that utilizes a comparator. To accomplish digital communication, the bit stream to be sent is modulated onto a carrier signal and transmitted. One of the most basic and widely used schemes is Binary Phase Shift Keying (BPSK). In BPSK, the amplitude of the carrier is constant, and the phase is shifted by +90° for a "one" bit, and −90° for a "zero" bit.

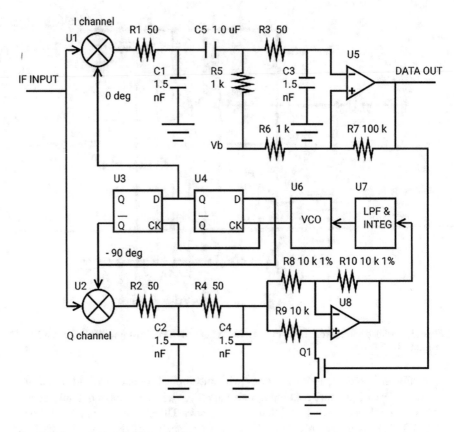

Fig. 5.4 Costas Loop BPSK demodulator

It turns out that this modulation totally suppresses the energy at the carrier frequency, requiring that the receiver either perform a squaring operation on the signal to recover the carrier for synchronous detection of the modulation, or a more ideal method is the Costas Loop, a special form of phase-locked loop. That is the example chosen here to illustrate this important use of the comparator in recovering the modulating bit stream. Reference: Wikipedia (2021).

While the circuit in Fig. 5.4 is arranged for demodulating BPSK, a relatively minor addition to this circuit enables it to demodulate Quadrature Phase Shift Keying (QPSK), where the I channel and the Q channel each independently carries bit streams. QPSK, therefore, carries two bits on information per symbol of modulation.

In Fig. 5.4 there is a phase-locked loop (PLL) to track the carrier and regenerate it as a reference for synchronous demodulation. The PLL is comprised of U2, U8, U7, U6, U4, and U3. U2 is the phase detector, and because the tracking sense of this loop shifts between negative and positive feedback due to the modulation phase shifts, it is necessary to invert the sense of the loop, or not, depending on the state of the modulation. This modulation-dependent inversion is accomplished with Q1 and an operational amplifier, U8. The output of U8 is a modulation corrected phase error that is fed to U7 for filtering and integration to set the tracking bandwidth of the PLL. The output of U7 is the control voltage that drives the Voltage Controlled Oscillator (VCO) U6 to determine its frequency. The VCO runs at four times the carrier frequency to drive a Johnson counter composed of U3 and U4, both D-type flip-flops. The Johnson counter divides the VCO output by four and generates an in-phase reference signal for the I Channel, and a very accurately 90° lagging (Quadrature) reference signal for the Q Channel. That completes the path around the PLL and the details will be discussed later.

The I Channel is where the data demodulation occurs as the mixer U1 is fed with the 0°, in-phase, carrier reference from the PLL. The output of U1 contains the bit stream, as band-limited by the earlier stages of the receiver (not shown) and the noise and interference encountered in the transmission path. This analog signal is filtered further and presented to U5, the comparator. This comparator, U5, then makes a hard decision as to whether the presently demodulated level represents a "one" or a "zero," resulting in the Data Out signal. The common-mode voltage for the comparator, as well as the local decision level, is set by the bias voltage Vb.

Should it be anticipated that the incoming signal level will extend over a wide range, it may be well to consider inserting the operational amplifier with soft-limiting diodes in the feedback path that is shown in Fig. 5.1. The opportune place in Fig. 5.4 to add this stage is after R1 and C1, and before C5.

In Fig. 5.4 there are many passive component values given to help illustrate some of the issues that must be addressed in such a design. Starting with the low-pass filter, R1 and C1, in the I Channel, those values were chosen for the following reasons: The mixer U1 might be a passive double-balanced mixer that is expecting to see a 50 Ω load, hence the value of R1. The mixers U1 and U2 could just as well be any other analog multiplicative element, such as a Gilbert cell multiplier.

With R1 chosen, the value of C1 should depend on the frequencies present. For this example, an IF Input frequency of 10 MHz was chosen, along with a data rate of 1 Mbps. The output of U1 will be the bit stream (and noise) plus a strong component of twice the IF and reference frequency, or 20 MHz. Before presenting this signal to the comparator, it is necessary to attenuate the 2IF component, while avoiding attenuation of the components of the data rate. With a data rate of 1 Mbps, the highest fundamental frequency will be 500 kHz (with harmonics). For these reasons, a corner frequency for R1C1 was chosen to be 2 MHz, because it gives 20 dB of attenuation at 20 MHz, and very little attenuation at 500 kHz. Since 20 dB of attenuation of the 20 MHz component is likely inadequate, a second section of this low-pass is added at R3C3, giving a total of 40 dB attenuation.

This design makes the simple assumption that the mean of the recovered data plus noise is the best decision level for the comparator, U5, and that is the function of C5 with R5. This is an implementation of the arrangement shown in Fig. 3.3, under Decision-Level Setting—Adaptive Thresholds, Average Tracking, and the discussion there applies. The averaging time constant of this high-pass filter must be chosen, and in this case, the value for R5 was set by concerns in the next paragraph.

That left C5 as the component that determines the averaging time, which in this case was somewhat arbitrarily set to 1.0 ms, being 1000 bit periods of the data to minimize data dependency of the threshold, but still short enough to handle propagation shifts in the path. In a system design, it is necessary to consider what the longest run of zeros or ones might be and keep this averaging time significantly longer.

As mentioned above, the value of R5 was set by another issue, that being a desire to keep the dc path to the two inputs of the comparator to have the same resistance. This keeps the bias current of the comparator from developing an additional offset voltage. That made R5 need to equal R6. The value for R6 at 1 kΩ was set based on wanting this stage to have 1% hysteresis, and the largest value that was desired for R7 was 100 kΩ.

It is worth mentioning that the reason the R5C5 high-pass was inserted between the two low-pass sections R1C1 and R3C3 was because it was desired to have R1 terminate the mixer U1, and it is desirable to have C3 near the input of U5 to improve its switching performance.

In the Q Channel, it may seem that R2C2 and R4C4 are superfluous, but in fact they are present for an important purpose. In the I Channel, the low-pass filters R1C1 and R3C3 insert a group delay in that signal path. It is necessary to have a very similar delay in the Q Channel in order to have the error signal arrive at U8 and Q1 at the same time as the data is applied to Q1 to make the modulation dependent inversion. In this inversion stage, note that R8 and R10 are shown as 1% tolerance parts, while R9 is not. This is necessary to keep unity gain in this stage for both states of inversion or not. It turns out that the value of R9 is not at all critical to the functioning of this stage, as long as the on-resistance of Q1 is much smaller, and the associated capacitive reactances are much larger.

5.4   Multiple-Comparator Circuits

5.4.1   Window Comparator

Two or more comparators can be arranged to monitor a voltage level, signaling whenever the level goes above or below preset limits. Fig. 5.5 shows a two comparator circuit that connects the open-collector outputs of a part like the LM393 to form a wired-OR such that if either limit is exceeded, the output drops. A pull-up resistor, R4, is needed at the output.

Here the voltage divider chain R1, R2, R3 set the reference levels for the two comparators. We may call the voltage at the R1, R2 node V1 and the voltage at the R2, R3 node V2. Then the output is high only if (Vin < V1) AND (Vin > V2). If the (+) and (−) inputs of the two comparators are swapped, then one of the comparator outputs will always be low, and the circuit is useless. If hysteresis is needed, it may be added to the window comparator by connecting large resistors from the output to

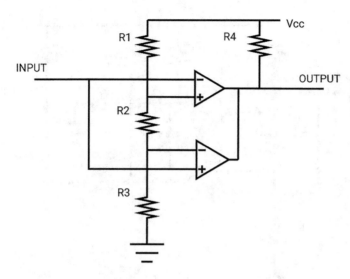

Fig. 5.5 Two comparators connected as a window comparator

the (+) inputs. For the lower comparator, a small resistor will need to be added in series from the input to the comparator (+) terminal.

Another connection of two comparators to establish a window determination displayed by three LEDs can be done very simply by connecting three LEDs among the open-collector outputs, which will then show High, OK, Low, as in Fig. 5.5. Once again, this circuit uses comparators having open collector (or open drain) outputs.

The function of the circuit in Fig. 5.6 is to compare the input voltage Vi with the two levels, V1 and V2, set by the resistive divider R1, R2, R3, and to light only one of the three LEDs at a time. The operation is that if Vi >V1 then D1 lights (High). If V1>Vi>V2 then D2 lights (OK), and if Vi <V2 then D3 lights (Low). While this is a clever arrangement, it has the disadvantage of drawing maximum current through R4, R5, and R6 whenever Vi < V2.

There is an alternative to the circuit in Fig. 5.6 that avoids the wasteful supply drain, but requires that the comparators have outputs that can source as well as sink current. If speed of response is not an issue, as with LEDs for a human interface, then this application could be met with operational amplifiers.

The window comparator in Fig. 5.7 operates in a similar fashion to the one in Fig. 5.6, but it avoids the high power drain state, and its different topology may appear somewhat "upside-down." The current sourcing specification for the comparator outputs needs to support the desired current flow for the LEDs. Of course, the current sinking specification for the comparators must likewise support the LED currents.

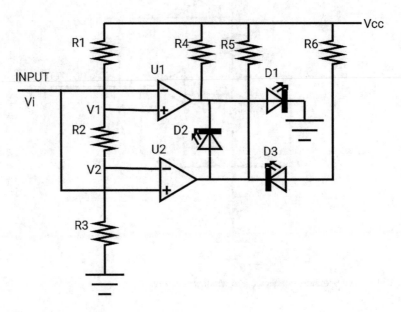

Fig. 5.6 A window comparator with LED indicators

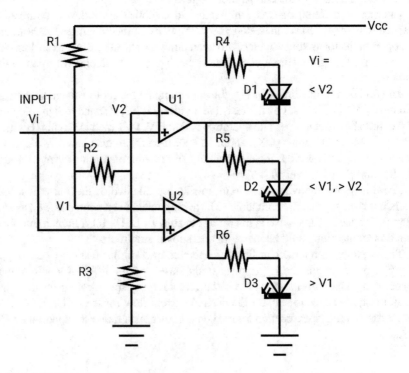

Fig. 5.7 Window comparator with LEDs, lower power

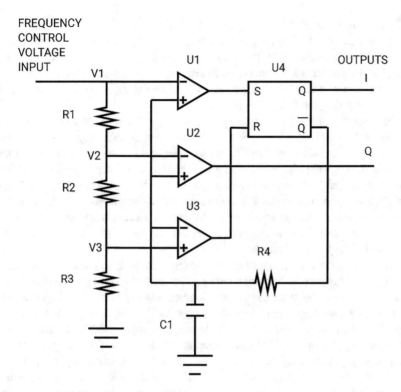

Fig. 5.8 Voltage-controlled oscillator with I and Q outputs

5.4.2   *Voltage-Controlled Oscillator*

Two comparators may be combined to make a voltage-controlled oscillator, VCO, and a third comparator may be added to provide a quadrature-phase output. This is detailed in Fig. 5.8. Reference Texas Instruments (2014)

The operation of this circuit can be described as a relaxation oscillator, but the details of that operation show that its functioning can be rather precise. To follow the sequence of events in the circuit, start with a power-on condition of little charge in C1, and a significantly higher control voltage V1 at the input. This would mean that the comparator U1 would see a higher voltage on its negative input, making its output low. Assume the SR flip-flop U4 is in the reset state, meaning its/Q output is high. This condition charges the timing capacitor C1, and its voltage ramps up. Since the voltage on C1 is also less than V2, then the output of U2, OUTPUT Q is also low. About halfway through the charging of C1, its voltage reaches V2, and the output of U2 OUTPUT Q goes high.

When the voltage on C1 reaches the control voltage V1 on the input, the comparator U1 changes state with its output going high, which activates the S or set input on the flip-flop U4 causing it to change to a set state. This drops the/Q output of U4 and begins to discharge C1. Since C1 has reached V1, then it is higher than V2, and the output of U2 is high at OUTPUT Q. As C1 discharges, when it reaches V2, the output of U2, OUTPUT Q goes low. This can be near exactly halfway in the discharge time of C1. As C1 continues to discharge, it eventually reaches V3. At this point, the output of U3, which has been low, changes to high, and this takes U4 to a reset state, starting the cycle of events over.

The accuracy of the 90° phase angle between the I and Q outputs depends on a number of factors: One is having R1 and R2 be equal. Another factor that can affect this accuracy is the differing input offset voltages of the three comparators. Another is having the control voltage be near half of the Vcc, or near the middle of the logic swing out of the/Q port of the flip-flop U4. This is due to the capacitor's charging curve being exponential rather than linear. An improvement would be to replace R4 with a switched current source that would make C1's charging linear with time and no longer dependent on the relationship of the control voltage to the logic output swing and its exponential character.

Another approach to linearizing the charging process is to place an integrator in the R4 path. This may be implemented as an op-amp in the R4 path, with C1 being the feedback capacitor from the output of the op-amp to its negative input. R4 now connects to the op-amp negative input, along with C1. The positive input of the op-amp should be biased at the midpoint of the logic swings from/Q. The output of the op-amp drives the inputs of the three comparators as did the C1 voltage before.

Another point that may be less than obvious is that the frequency versus control voltage function is inverse or negative. That is to say that as the control voltage increases, the frequency decreases. This is because the capacitor charges at a rate that is not affected by the control voltage, and a higher control voltage causes the capacitor to charge for a longer time to reach the higher voltage.

A similar circuit arrangement is used in the LM567 VCO integrated circuit where only the capacitor C1 and its charging resistor R4 are external to the chip and amenable to selection for performance.

Should the quadrature output of this VCO be unneeded, then U2 may be omitted, and R1 and R2 combined.

5.4.3   Dot-Bar Display Drivers

At one time there was a family of dot-bar display driver integrated circuits with multiple comparators to drive LED indicators. There were three versions; the LM3914 with a linear scale, the LM3915 with a decibel scale in 3 dB steps, and the LM3916 with a scale similar to a VU meter. These were National Semiconductor products and since Texas Instruments has acquired National, the −15 and −16 have been discontinued by TI, and no replacements are known. The currently available product, LM3914, with 10 steps, may be cascaded up to 10 units for 100 steps; and its data sheet has a large number of application details. The cascaded operation allows either dot mode or bar mode and there is also a zero-center connection. The outputs can drive LEDs, LCDs, or vacuum fluorescents. Figs. 5.9, 5.10, and 5.11

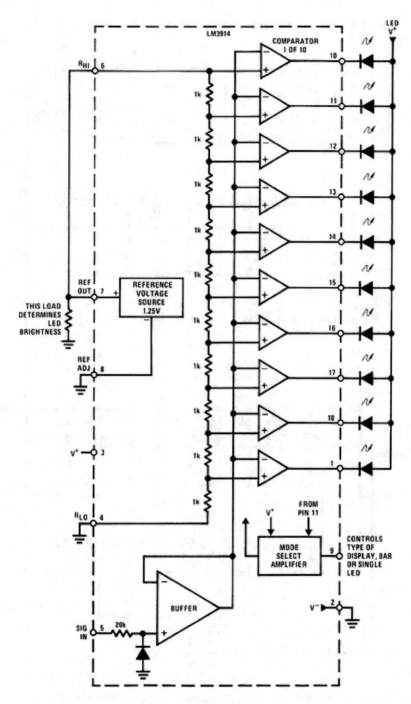

Fig. 5.9 LM3914 block diagram; simplest application (Courtesy Texas Instruments)

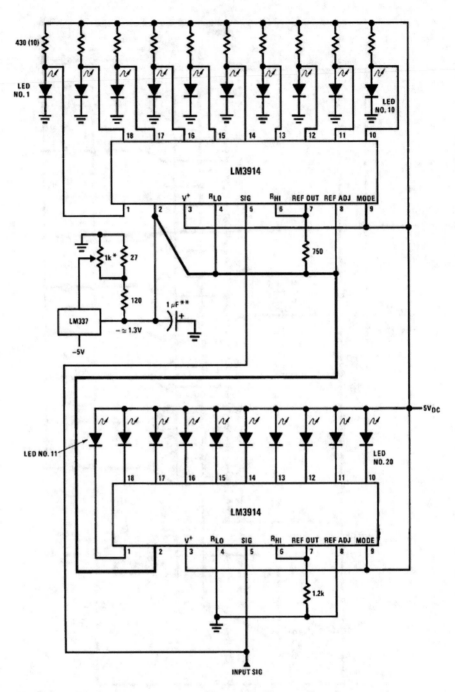

Fig. 5.10 LM3914 as a zero-center meter, 20 segments (Courtesy Texas Instruments)

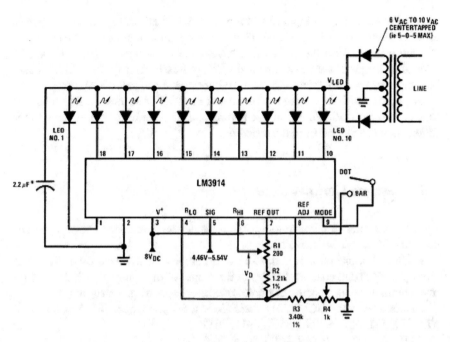

Fig. 5.11 LM3914 with expanded scale (Courtesy Texas Instruments)

show some of the connections and arrangements, including the simplest, zero center 20 segments, and expanded scale. Reference Texas Instruments (2013)

The LM3914 can be connected to have various step increments between the segments. In the simplest connection, the steps are 125 mV. It can be connected for as little as 20 mV/step in bar mode and 50 mV/step in dot mode.

In Fig. 5.11, the voltage Vd is adjusted to give the display the desired full-scale range. (Vd is from the slider on R1 to the opposite end of R2.)

5.4.4 Flash Analog-to-Digital Converter

The flash analog-to-digital converter (ADC) is an extension of the window comparator above. A single comparator resolves an analog signal into a single bit. Two comparators can resolve an analog signal into three levels, which require two bits to represent, while one of the four states of the two bits is never used. Three comparators can resolve an analog signal into four levels, which can be encoded as two bits. Similarly, seven comparators can resolve an analog signal into eight levels, encoded as three bits; and 2^{N-1} comparators can resolve to 2^N levels, encoded as N bits, with 63 comparators yielding 6 bits, or 255 comparators yielding 8 bits, for example.

The circuit configuration of a flash ADC is an extension of the window comparator, in that a resistive divider string has taps at each of the level transition edges, each tap feeding the (−) input of a comparator, while all the (+) inputs are ganged together being fed the input signal. Then the comparator outputs must be combined in logic gates to form the output bits. A version of this logic function is called a "Wallace Encoder," which outputs a simple progressive binary code. Other codes that might be used are "Gray Code" or "binary-coded decimal (BCD) code."

The name "flash ADC" comes from the high speed of the device. Its conversion time is the propagation delay of one comparator plus the propagation delay of the following gates that perform the encoding.

5.5 Latched Comparators

Specific examples of comparators without latches were included above in the discussion of AM-to-PM Conversion.

A latched comparator is composed of a comparator followed by a flipflop. A clock pulse or latch enables samples of the output of the comparator and stores the result in the flipflop. Specific examples of comparators with latches are:

Analog Devices: ADCMP551, 552, 553, 561, 562, 563, 564, 565, 566, 567, 572, 573, 580, 581, 582, AD790, 8561, 8611, 8612.

AMD: Am685, 686, 687, 1500, 6685, 6687.

Hittite: HMC674LC3C.

Maxim: MAX9600, 9601, 9602.

National Semiconductor: LMH7322.

Texas Instruments: TL3016.

Since the output of a latched comparator will generally follow the enable or clock pulse, the hysteresis schemes shown above are not applicable.

Many of the highest performance (fastest, widest bandwidth) comparators are equipped with a latch (flip-flop) at their output. Most of these latched comparators have a logic level input called latch enable. Whenever the latch enable is true, the comparator responds as if there were no latch. When the latch enable goes false, the output state is held at its last state. This latch enable can be employed to implement a sampling function, so that the output reflects the condition existing at the instant the latch enable went false.

Some of the latched comparators have the latch operating in the mode of a type D flip-flop, where the latch is updated in response to a clock input. The clock input to the latch accomplishes a sampling function, allowing the output to ignore changes in the input, reflecting only the condition existing at the instant of the clock pulse. Further, the output can only change when a clock pulse occurs. This clearly avoids changes in the comparator output between clock pulses and forces the output data stream to be synchronous with the clocking function.

References

Comparator circuit details—Texas Instruments (2013) data sheet LMH7322

Costas Loop Demodulator, BPSK, QPSK—Wikipedia (2021) https://en.wikipedia.org/wiki/ Costas_loop. Accessed 2021 Oct 20

Voltage Controlled Oscillator—Texas Instruments (2014) data sheet LM567, p 8

Dot-Bar Display Drivers—Texas Instruments (2013) data sheet LM3914

Latched Comparators—Data sheets of the listed parts n.d.

Chapter 6
Logic Elements

6.1 Logic Elements as Comparators

While digital logic functions are not characterized as comparators (with a few exceptions in the emitter-coupled logic [ECL] families), they are sometimes the hardware of choice for what is otherwise a comparator function. Features of interest in this category are the fact that the threshold is built into the structure of the gate and not accessible on a pin, the threshold is very loosely specified and usually has significantly more temperature sensitivity than comparators as a group, most logic families are noisy in the sense of random noise present at the input, and three different levels of internal hysteresis are available: none, small, and large.

The loosely specified threshold level and its variation with temperature can be reduced by negative feedback via an integrator. Note that the combination of having both hysteresis and integrated negative feedback makes the slicer stage oscillate at a fairly low frequency in the absence of an input signal. This will be analyzed in more detail later in this chapter.

What conditions might suggest the use of a logic element as a comparator? If the analog signal is robust and comes from a relatively low impedance source, that's 0.5 Vrms or 1.4 Vpp from a source of <1 kΩ, and it doesn't change much from max to min, like 20% total, then a transistor–transistor logic (TTL) gate input may suffice to digitize it. A CMOS gate can have a somewhat more tolerant input and with its very high input impedance >1 MΩ, it can function well with higher impedance circuitry at its input. An example of this is a switch de-bounce circuit, discussed earlier. An ECL gate input can work with smaller signal swings. But all of these suffer from more or less temperature sensitivity to changes in the reference level and changes in hysteresis, if present. And if dispersion is a concern (discussed in Dispersion, AM-to-PM Conversion), logic elements offer no characterization of this behavior.

Early in the history of integrated logic, ECL held a performance edge over TTL in toggle rate and propagation delay, mainly because the transistors in ECL were

© The Author(s), under exclusive license to Springer Nature Switzerland AG 2022
M. C. Fischer, *Comparators*, https://doi.org/10.1007/978-3-030-95742-1_6

never operated in saturation, while TTL did. Advances in TTL, namely Schottky versions that avoided saturation, narrowed this gap, and more recently, CMOS and low voltage logic have competed well with ECL.

The structure of ECL is fundamentally quite similar to that of Fig. 1.3, being entirely NPN transistors. While most logic functions in ECL have one of the differential inputs connected internally to a bias voltage, the data receiver units have both inputs accessible. An example part is the 10EL16/100EL16. References Texas Instruments (2021), OnSemi (2021).

Original ECL families were specified to have a negative supply of −5.2 Vdc, with the positive supply being ground. ECL may also be used with a +5 Vdc supply, with the negative supply being ground, in which case it is called PECL (Positive Emitter Coupled Logic), and sometimes the negative supply connection is called NECL.

ECL inputs are fairly high impedance, with moderately low bias current. The EL16 parts specify an Input HIGH Current maximum of 150 µA. For a positive emitter-coupled logic (PECL) +4 V high level, this represents a 27 kΩ static resistance. Many families, like the 10EL/100EL, have a nominal 50 kΩ pull-down resistor to the negative supply on each input, but this value ranges widely, is not specified, and may be inferred only from the high-level input current, as in this case, which undoubtedly includes the input transistor base current. The 10EL parts have the temperature variations of the earlier versions of ECL, while the 100EL parts are temperature compensated with a much-improved characteristic.

There are now low-voltage versions of ECL that operate from 3.2 Vdc. An example part with a comparator offering both differential inputs is the 10EPT21 (also 100EPT21 and 65EPT21 as well as 10/100LVEL16). These inputs have 50 kΩ pull-up resistors and 50 kΩ pull-down resistors, giving the part a nominal 25 kΩ input impedance, plus transistor base current. This is only specified at PECL logic levels as a maximum input current of ±150 µA. Another part in this class is the 9321B that operates from supplies between 3.0 and 5.5 Vdc, either PECL or negative emitter-coupled logic (NECL), with response to 3 GHz (for data rates of potentially 6 Gbps).

Some ECL circuits, like the ones mentioned here, have a built-in bias voltage source that is a desirable common-mode level for the differential inputs. This bias voltage is appropriate to connect in place of the ground symbol shown in Figs. 3.1, 3.3, 3.4, 3.14, and 3.15 and for the Vb shown in Fig. 3.6.

TTL may be thought of as a common-emitter connected amplifier, where the threshold voltage is the sum of two Vbe voltages, with their attendant significant temperature sensitivity. TTL inputs are normally moderately high impedance, with a significant bias current, of a negative polarity, that is, flowing out of the input.

CMOS logic is offered in two families that have a threshold voltage that is either approximately half the supply voltage (as in 74HC), or offset from that value to more closely approximate TTL, for ease of interfacing (as in 74HCT). CMOS inputs are normally very high impedance.

Most logic families have no hysteresis built in. Exceptions are the CMOS 74AC and 74ACT which have a small amount of hysteresis built into each input of approximately 100 mV to improve noise margins and to avoid oscillation with input transitions of ±10 ns/V or faster. [ref: Texas Instruments (1987)] Further, in both 74AC and 74ACT, Texas Instruments and Philips/Signetics have included a technique called dynamic hysteresis which provides a significant amount of hysteresis. (ibid. p. 3–25) The range of this hysteresis is evident from the data sheets of the specific parts. And the Texas Instruments *AHC/AHCT Designer's Guide* (SCLA013), page 10 mentions that a kind of Schmitt-trigger circuit is part of the input stage. The hysteresis of this is about 200 mV. Also, see the section below on "Logic Elements' Thresholds and Hysteresis."

Large amounts of hysteresis are built into the gates and buffers listed as "Schmitt" devices, such as 74AC14. Since logic gates have only a single input with the threshold set internally, then the AC coupling mentioned earlier is particularly appropriate. The DC return of the shunt resistor in the coupling network must be fed with a voltage that approximates the threshold of the logic, unless an offset threshold is desired.

A non-inverting logic element with zero hysteresis can be made to operate with hysteresis by connecting feedback from its output to its input (positive feedback), and since this is done with external components (resistors), it is adjustable and fairly stable. This technique is discussed in more detail in the paragraphs associated with Fig. 6.3.

With low slew rate, that is less than 0.10 V/ns, the hysteresis of a Schmitt trigger is required to avoid oscillations at the transition with AHC. Slew rate is discussed above in Hysteresis.

The dispersion of logic elements used as slicers is unknown, unspecified, and unpredictable. If dispersion is an issue, then a logic element should be avoided.

When logic families are evaluated for the jitter that they add to a signal passing through them, this is almost always done with inputs that are typical output waveforms for the logic being evaluated, and are therefore high-level, high slew rate digital waveforms. This gives a result that is minimally perturbed by the equivalent input noise as was discussed earlier in the section Input Noise Measurement. An input with a lower slew rate could incur orders of magnitude more jitter.

The equivalent input noise of a number of logic families has been evaluated and the results are summarized in Table 6.1, where it is seen that the Schottky families, as in 74LS, can have good performance. However, the best performance is seen where particular attention has been paid to minimization of jitter in CMOS and ECL families.

Reference: Analog Devices (2008)

A logic element operating as a slicer may be used to determine the zero crossings of AC power line voltage. This has been done successfully by using large series resistors with a shunt capacitor, and shunt clipping diodes, into a Schmitt trigger buffer, as in Fig. 6.1. This circuit operates over line voltages from 90 to 270 VAC, 50 and 60 Hz.

The components in Fig. 6.1 are described rather more specifically than the components in most of the other circuits, and that is because of the specific application where the input voltage range, the frequency range, and the application system, a smart electric watt-hour meter, is known. The input resistor is split into two

Table 6.1 Summary of Logic Families and Their Additive Jitter

Logic family	Jitter; comments
FPGA	33–50 ps
(driver gates only, not including internal DLL/PLL) (1)	
74LS00	4.94 ps (2)
74HC700	2.2 ps (2)
74ACT100	0.99 ps (2)
MC10EL16 PECL	0.7 ps (1)
AD951x family	0.22 ps (1)
NBSG16 Low Swing ECL (0.1 V)	0.2 ps (1)
ADCLK9xx ECL Clock Drivers	0.1 ps (1)

(1) Manufacturer's specification
(2) Calculated value based on degradation of ADC SNR

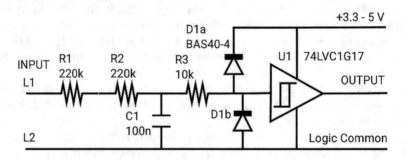

Fig. 6.1 Power line zero-crossing detector

components, not because of power dissipation (a maximum of about 77 mW each) but because of the need to survive much more extreme power line transient voltages. These sharp transients are also one of the reasons for C1, in addition to providing a low-pass filtering function to attenuate interfering frequencies above 60 Hz (AM broadcast, power line communication signals, switching power supplies, etc.) Since U1 is a Schmitt trigger logic buffer, the shape inside the symbol U1 is the usual indication for a Schmitt trigger device, a comparator with hysteresis. The shape of the marking comes from a plot of voltage out versus voltage in, showing the hysteresis.

Since the diodes in the dual Schottky diode assembly, D1, are conducting most of the time, R3 provides a fixed impedance for C1 to work against in the low-pass function. The diodes, D1, are chosen to be Schottky barrier, with their low forward voltage, to take most of the excess input signal current rather than relying on the input ESD protection inside the Schmitt trigger buffer, even though it is rated to 50 mA.

Clearly, the threshold of U1 is near half the logic supply voltage (as discussed later in the Sec 6.3 Logic Elements' Thresholds and Hysteresis). This represents an offset of a couple of volts from the actual zero crossing of the input waveform. Since the input waveform is at least 100 Vac, with a peak value of 141 V, then this offset represents just over 1% of the peak value of the waveform, generally negligible. If this offset is undesirable, then a series capacitor can be inserted at the input before R1. The reactance of this capacitor should be a fraction of the sum of R1+R2 to minimize the phase shift it incurs. The added capacitor will center the waveform on the threshold of the logic element, due to the action of D1 and D2, to within a fraction of a volt. By setting the value of the added capacitor appropriately, the phase lead caused by this capacitor may be adjusted to compensate for the phase lag due to the hysteresis of U1.

Since the circuit in Fig. 6.1 is shown being connected directly across the AC power main, it is imperative that proper safety precautions be observed when working with such dangerous and potentially lethal conditions.

6.2   Logic Elements as Oscillators

As was mentioned earlier, an inverting comparator with hysteresis and negative feedback will oscillate. In logic families, the Schmitt trigger function provides the hysteresis, and an inverter with the addition of a resistor and capacitor as negative feedback, results in a crude resistance–capacitance (RC) oscillator, as in Fig. 6.2.

While being the simplest of oscillators, this should only be used where the resulting frequency is not at all critical. There is a formula for estimating the frequency:

For 74AHC14: $f = 1/T \sim 1/(0.55*R*C)$

The approximate nature of this formula may be inferred from the range of the hysteresis band. The hysteresis of the 74AHC14 is specified to range between 0.4 and 1.4 V, and the resulting oscillation frequency will cover a proportionately broad range (±56% of nominal). In this component, the hysteresis varies mainly on a unit-to-unit basis, with a smaller effect of supply voltage, and an even smaller effect of temperature. These points would suggest making R1 variable, to be set in final test,

Fig. 6.2 Schmitt inverter
RC oscillator

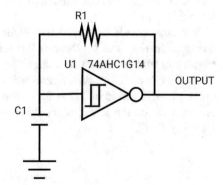

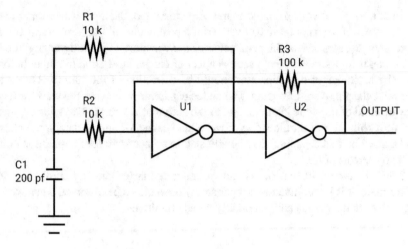

Fig. 6.3 RC oscillator made with non-hysteresis inverters

to bring the frequency into a required band. Even so, this is still one of the most crude ways of generating an oscillating signal, while also being one of the simplest.

With two non-hysteresis logic inverters, a version of the above circuit may be realized with more predictable frequency characteristics. There are two criteria that must be satisfied to make this work: First, hysteresis must be functional in an inverting stage, and second, a non-inverting path must be available to provide the positive feedback required to accomplish hysteresis.

This is detailed in Fig. 6.3 where the second inverter provides the positive feedback allowing hysteresis to be applied through R3 working against R2. The amount of hysteresis is the U2 output logic swing attenuated by the voltage divider R3 and R2. In a system where a low-precision clock is needed, and there are no Schmitt inverters but there are spare inverters with no hysteresis, this circuit can serve.

With the component values shown in Fig. 6.3, the operating frequency should be near 1.0 MHz. This is more predictable than the circuit in Fig. 6.2 because, regardless of the supply voltage, the amount of hysteresis is set primarily by the ratio of R3 and R2.

The capacitor C1 is charged and discharged over the hysteresis range by the logic swing of the output of U1 through R1. With the component values shown, the hysteresis band will be about one-tenth of the logic swing. The time constant of R1 with C1 is 2 us, and the hysteresis band will be covered in about 0.5 us, for each half-cycle, giving a period of 1.0 us for the output square wave, or 1.0 MHz. Of course, this depends on the actual threshold level of the input of U1, which for CMOS is typically Vcc/2.

Table 6.2 Threshold voltages of TTL/CMOS logic technologies

Threshold voltage	Device technology; Vcc
1.5 V	ABT, AHCT, HCT, ACT, ALS, AS, F, LS, S, TTL (all $Vcc = 5$ V)
$Vcc/2$	HC, AC ($Vcc = 2$–6 V)
	AHC, LV-A ($Vcc = 2$–5.5 V)
	LVC, ALVC, AVC ($Vcc = 1.7$–3.6 V)
	LVC1G/2G/3G ($Vcc = 1.65$–5.5 V)
	AUC ($Vcc = 0.8$–2.7 V)

6.3 Logic Elements' Thresholds and Hysteresis

Since many applications of logic elements as data slicers may be met by one or another type of buffer, that is the focus of this discussion. As is familiar in the areas of digital logic integrated circuits, the following Table 6.2 lists logic functional types as their numeric designators, e.g., 07, 4049, and the families of supply voltage and performance as their alphabetic designators, e.g., HC, AVC.

First the TTL/CMOS families:

- Buffers:

  - Non-inverting: 07, 17
  - Inverting: 04, 06, 14, 16, 4009, 4049

Buffers without hysteresis: 04, 06, 07, 4009, 4049.
Buffers with small hysteresis: AC, ACT, AHC, AHCT needs $dV/dt \geq 0.10$ V/ns.
Buffers with large hysteresis: LVC14, LVC17, Vhyst = 0.40–1.1 V.
Next, the ECL families:

For frequencies reaching 1.0 GHz and above, the emitter-coupled logic (ECL) components may be required. In the ECL families, the comparator functions are often called, "Differential Receivers," grouped in categories of, "Drivers and Buffers." The basic numeric designations are 11, 16, 21, and 23. The MC10EL16 is a representative component. The MC100EL16 version is the same except that the 100 family is temperature compensated, likely a benefit to most comparator applications. The term, "differential," indicates that both inverting and non-inverting inputs are accessible, making these logic elements applicable to many of the circuit arrangements discussed here. Note that many ECL comparators (buffers) have pull-down resistors attached to their inputs, typically 75 kΩ to the negative supply, VEE. While still a relatively high impedance, this may need to be considered in the design of the driving circuitry. Of similar concern is the fact that the input bias current can be as high as 150 μA. ECL comparators generally do not have hysteresis built in, often have a threshold voltage supply pin, and have no specification of dispersion, the variation of propagation delay versus overdrive, nor equivalent input noise.

Where a crude comparator function with hysteresis is needed, and two inverters are available, the circuit of Fig. 6.3 provides an alternative by removing C1 and R1, and driving the left end of R2. The input signal is connected in place of C1, likely through a capacitor. If this stage is driven through a capacitor, then it operates in the mode of average tracking adaptive threshold, discussed earlier, with the averaging time constant of $\tau = C1(R2+R3)$. The driving stage needs to have an output impedance much less than R2, or otherwise that driving impedance must be added to R2 to determine the amount of hysteresis. With a low driving impedance, the hysteresis of such a stage is set by the ratio of R1/R2, and can be rather precise and stable. What remain as crude are the input noise, the threshold voltage, and temperature effects. Of course, this approach also works with a single non-inverting buffer, which combines the function of U1 and U2, with all the same advantages and disadvantages.

References

Analog Devices (2008) Analog-Dialog http://www.analog.com/en/analog-dialogue/articles/analog-to-digital-converter-clock-optimization.html. Accessed 2021 Oct 17

Texas Instruments (1987) Advanced CMOS Logic Designer's Handbook, SCAA001, 1987, p 3–12

Texas Instruments, PECL/ECL Buffers and Translators http://www.ti.com/download/aap/pdf/PECL-ECL_Buffers_and_Translators.pdf. Accessed 2021 Oct 19

OnSemi (2021) Drivers and Fanout Buffers http://www.onsemi.com/PowerSolutions/parametrics.do?id=547. Accessed 2021 Oct 23

Chapter 7
Comparator Design and Effects

7.1 Bipolar Integrated Comparator Design

In the much earlier section, Basic Concepts, Figs. 1.3 and 1.4 detail simple circuits realizing the comparator function in bipolar technology, along with a discussion of some of the considerations in the design of Fig. 1.4. Basically a bipolar comparator design starts out along the same lines as a bipolar operational amplifier. Then after the input stage, instead of designing for internal compensation to allow closed-loop amplification, there is no internal compensation, rather the concern is with avoiding saturation of all stages that would slow the response. The basics of bipolar comparator design are covered along with some clever advances in reference Hamilton (1975).

7.2 MOS-Integrated Comparator Design

Designing a comparator utilizing MOS transistors is different from the bipolar case in that saturation of the transistors is not a concern. That means that the design may proceed along the lines of an operational amplifier, with concern for the usual case that the output levels should be compatible with the high and low logic levels of whichever logic family with which the comparator is intended to be used. The Wikipedia Comparator reference shows a low-power CMOS dynamic latched comparator. References: Wikipedia Comparator (2021), Rahman (2014).

It would be tempting to consider the CD4007 as a candidate for breadboarding an MOS comparator, but some of the six transistors in that device are connected internally making them either inconvenient or inaccessible to this purpose. I have not found any way to use a CD4007 alone to breadboard a comparator with any reasonable characteristics.

7.3 Choices of Semiconductor Technology

There are two areas of technology choices to be considered. The first is the semiconducting device material. Some alternatives are silicon, silicon–germanium, silicon carbide, gallium arsenide, and gallium nitride. Silicon is the most widely used, while silicon carbide is seeing increasing use for high-temperature applications. The highest speed, highest frequency applications are accomplished with silicon–germanium and gallium arsenide.

The newest work is being done with gallium nitride, first used for blue light–emitting diodes, at present showing great promise in radio frequency (RF) amplifier transistors with power levels of tens to hundreds of watts at multi-gigahertz frequencies, but no low noise versions yet. The gallium nitride discrete transistors currently available tend to be field-effect transistors (FET) of either depletion-mode or enhancement-mode high-electron-mobility transistors (E-HEMT) having some similarities to silicon MOSFET devices. Their speed is enhanced by having no minority carriers and no body diode. Reference: Wikipedia Gallium nitride (2021).

The second choice to consider is the device structure, among bipolar junction transistors (BJT), junction field-effect transistors (JFET), and metal-oxide semiconductor field-effect transistors (MOSFET). Most of the basic functions of a comparator and its specifications are widely known and documented as to choice of technology. The two areas of performance that have received much less attention are dispersion and noise. For that reason, they are given more thorough coverage in Chap. 4 for device performance. Considering those performance parameters with respect to technology choice is covered next.

In Chap. 4, Sect. 4.1 Dispersion, there is a lengthy discussion of the AM-to-PM effects of comparator dispersion. All the higher performance comparators mentioned are silicon BJT technology, except for the highest performance unit, which employs BJTs that are silicon–germanium. In contrast, the unit at the bottom of the list, with the worst dispersion, is implemented with CMOS technology, while a midrange unit has MOSFET inputs and CMOS outputs.

In Chap. 4, Sect. 4.2 Noise Measurement, the effects of noise generated in the comparator are discussed and shown to be an area of concern. In the present state of device technology, BJTs have the advantage for low noise contribution, while offering wide bandwidth and high speed. JFETs can approach bipolar noise levels but only with compromises of their other characteristics. Finally, MOSFETs (including complementary metal-oxide semiconductor CMOS) have not been able to reach the low noise levels of JFETs.

Besides the two choices of device material and device structure, there is likely to be a choice of technology, and it is at this point where the duality of the comparator function is significant. This duality is that the input structure of a comparator is analog while the output structure is digital, forcing such designs to be mixed signal. Whereas semiconductor process technology can be optimized for either digital or analog, in the case of the comparator, the mixed-signal compromises to cover both functions are required. In the case of each design, the need for low offset and low

noise in the input and fast switching in the output must both be addressed.
Fortunately, the need for large gain-bandwidth product applies to both the input and
output stages.

7.4   Hard Limiter Effects on Signal-to-Noise Ratio

The effects of a comparator on signal-to-noise ratio (SNR) are generally discussed
considering the comparator to be a hard limiter, a very good approximation.
Numerous authors have explored these effects, starting with Davenport in 1953. For
a single signal plus noise, the result may be summarized as the ratio of output SNR
to input SNR, which for positive SNR asymptotes to a factor of 2 for a gain of 3 dB,
while for a negative SNR it asymptotes to pi/4 which is near −1 dB. In the vicinity
of 0 dB SNR, the effect is much smaller. See the plot in Fig. 7.1, abstracted from
Comparatto (1989), based on work by Davenport (1953):

Note that for all positive signal-to-noise ratios, including 0 dB, the hard limiter
actually improves the system performance. And even for negative signal-to-noise
ratios, the penalty for hard limiting is at most 1 dB. While it may seem counter intui-
tive, this result is well established. Other treatments of this topic are found in
Blachman (1968), Chang (1970) and Middlestead (2017).

For a desired signal and an interfering signal plus noise, a range of these cases
was covered by Jones in 1963. Cases for multiple interfering signals were covered
by Jain in 1972, by Comparatto in 1989 and for some modulated signals by
Sevy in 1969

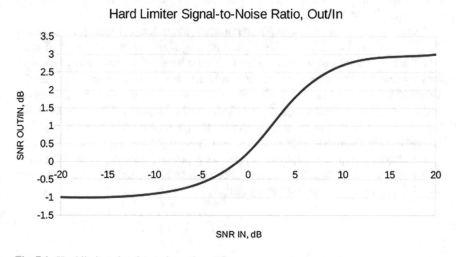

Fig. 7.1 Hard limiter signal-to-noise ratio, out/in

7.4.1   Benefits for a Phase-Locked Loop

A comparator can be very useful in the implementation of a phase-locked loop. When the loop is to lock to a variable and/or noisy analog signal as its input, performing an amplitude limiting operation just ahead of the phase detector can do two important things: (1) Amplitude variations are removed that could otherwise affect the transfer gain of the phase detector and therefore the loop bandwidth; and (2) the in-phase component of the noise is stripped off, improving the signal-to-noise ratio (SNR) by 3 dB. However, the effect of a better signal-to-noise ratio will not be reflected in the performance of the following loop, because the improved signal to noise ratio following the limiter is due to amplitude variations being stripped off the signal, and the phase information, including noise effects, is essentially unaffected by the limiting.

Further, if the SNR is negative, then the transfer gain of the phase detector is affected by variations in SNR, even with the amplitude limiter. This can be visualized by considering that the limiter output of noise is essentially constant for negative SNR, while the signal content of this output is falling with worsening SNR. This has the effect of narrowing the loop bandwidth as the SNR gets poorer, generally a desirable condition, except that it will likely increase the lock time and narrow the tracking range. Another view of this condition is that a negative SNR can "linearize" the result of slicing a signal that is changing slowly relative to noise with higher frequency components. In many cases, this reduction in transducer gain will be nearly linear, in that each decibel of worsening signal-to-noise ratio will result in about a dB of reduction of transducer gain.

These comments would apply to the detectors in Fig. 5.4, if a comparator (hard limiter) were the last stage of the IF chain, ahead of the detectors. This is not shown in Fig. 5.4, but would allow the detectors to be implemented with exclusive-or logic gates. This further interfaces well with the digital levels from the divider chain of the 0° and −90° reference signals.

Using a comparator for an amplitude limiter constitutes performing a "hard limiting" operation on the signal, with all the advantages of limiting in general. In addition, the comparator output, being standard logic levels, can drive a logic element for the phase detector operation. Alternatively, an analog phase detector is generally a double-balanced mixer, while the phase detection function for digital levels is performed by an exclusive-or gate (or a digital phase-frequency detector—see Table 6.1). A digital exclusive-or gate is almost always significantly smaller and less expensive than a double-balanced mixer, and there may be no performance penalty.

An alternative to the exclusive-or gate as a phase detector, the digital phase-frequency detector can exhibit faster, more reliable acquisition times for an input signal that occupies a wide frequency range.

7.4.2 Zonal Bandpass Filter

When there is noise or interfering signals present along with the desired signal, filtering is required. Typically a bandpass filter is desired, and if the aim is to attenuate signals much lower and much higher (as harmonics) than the desired signal, while allowing for modulation sidebands and variations in the desired signal frequency and variations in the tuning of the filter itself, then a zonal bandpass filter may be a good choice.

The term "zonal" means having a passband width approximately equal to the center frequency, as in a single tuned resonator with $Q = 1.0$, while the center frequency is the geometric mean of the two corner frequencies. A simple example is a pass band of $(2/3)f_o$ to $(3/2)f_o$, which has its geometric mean at f_o. In this case, the bandwidth is $(9/6) - (4/6) = 5/6$ which is a little less than f_o, having $Q = 1.2$. If the ratios of $(2/3)f_o$ to $(4/3)f_o$ are used, this filter has an octave of bandwidth, with a geometric mean (center frequency) of $0.889f_o$ and a $Q = 1.33$.

A single-tuned filter that has $Q = 1.0$ will have a pass band of $((1 + \text{sqrt}5)/2)f_o$ to $(2/(1 + \text{sqrt}5))f_o$, which reduces to $1.0f_o$. To four digits of precision, that upper pass band corner frequency is $1.618f_o$, and the lower is $0.618f_o$.

In rational terms, approximations are $13/8 = 1.625$ and $8/13 = 0.615$, which come closer than $3/2$ and $2/3$ to a $Q = 1.0$ alignment. [$8/5 = 1.600$, $5/8 = 0.625$; $17/10 = 1.700$, $10/17 = 0.588$; $19/11 = 1.727$, $11/19 = 0.579$]

This concept of a zonal bandpass filter may be best applicable after a superheterodyne down conversion (or possibly after a second conversion) where the desired signal is not yet demodulated, but the aim is to get the bandwidth narrow enough to achieve a positive signal-to-noise ratio.

If the desired signal is a stream of approximately random data bits, clearly that signal will occupy a bandwidth ranging from the data clock rate down to some quite low frequency with a period that is twice the longest run of zeros or ones to be expected.

All these points suggest the possibility of a data receiver that does down conversion to a moderate intermediate frequency, where the final filtering is approximately zonal, then feeding this signal to a comparator. The output of the comparator is then a bit stream of the carrier with its angular modulation, plus noise, and interference. This bit stream may then be processed digitally to do carrier tracking, data demodulation, and data clock recovery.

7.4.3 FM Capture Ratio

When two signals of similar frequency are passed by the bandpass filtering in the front end of an FM receiver, a phenomenon is observed that is called "capture." This occurs where, if one signal is slightly higher amplitude, that signal will suppress the propagation of the lower amplitude signal as they pass through the limiting stage.

This effect is known as "capturing the limiter," and is often specified as the minimum difference in the amplitudes of the two signals that yield a 20 dB difference in the demodulated results. That difference in RF amplitudes is called the "capture ratio" of the receiver and for well-designed circuits can be as little as 0.5 dB, thus yielding one of the most desirable features of FM transmission.

References

Hamilton, Douglass J., Howard, William G. (1975) Basic Integrated Circuit Engineering McGraw-Hill New York p 403-406

Rahman, Labonnah Farzana, Mamun Bin Ibne Reaz, Chia Chieu Yin, Mohammad Marufuzzaman, Mohammad Anisur Rahman (2014) A High-Speed and Low-Offset Dynamic Latch Comparator *The Scientific World Journal*, v 2014, Article 258068. https://doi.org/10.1155/2014/258068 Accessed 2021 Oct 31

Wikipedia Comparator (2021) https://en.wikipedia.org/wiki/Comparator Accessed 2021 Oct 31

Wikipedia Gallium nitride (2021) https://en.wikipedia.org/wiki/Gallium_nitride Accessed 2021 Oct 15

Davenport, W. B., Jr. (1953) Signal-to-Noise Ratios in Band-pass Limiters, _J. Appl. Phys._ 24, 720-7 June 1953

Jones, J. J. (1963) Hard-Limiting for Two Signals in Random Noise, _IEEE Transactions on Information Theory_, IT-9, p 34-42, January 1963.

Blachman, Nelson M. (1968) The Output Signal-to-Noise Ratio of a Bandpass Limiter, _IEEE Transactions on Aerospace and Electronic Systems_ Vol. AES-4, Nr. 4, July 1968

Sevy, J. L. (1969) The Effects of Limiting a Biphase or Quadriphase Signal Plus Interference, _IEEE Transactions on Aerospace and Electronic Systems_, Vol. 5, Nr. 3, May 1969

Chang, J.C. (1970) The Response of Hard-Limiting Bandpass Limiters to PM Signals, _IEEE Transactions on Aerospace and Electronic Systems_, AES-6(3):398 - 400 · June 1970

Jain, P. C. (1972) Limiting of Signals in Random Noise, _IEEE Transactions on Information Theory_, Vol. IT-18, No. 3, May 1972

Comparetto, G. M., & Ayers, D. R. (1989) An Analytic Expression for the Magnitudes of the Signal and Intermodulation Outputs of an Ideal Hard Limiter, Assuming n Input Signals Plus Gaussian Noise, International Journal of Satellite Communications, Vol. 7, 3-6 1989

Middlestead, Richard W. (2017) Digital Communications with Emphasis on Data Modems: Theory, Analysis ...", p 377, John Wiley, 2017.

Index

Printed in the United States
by Baker & Taylor Publisher Services